519.5
PUR

AF844739

180605056X
WITHDRAWN FROM STOCK

ack

12-00

Applied Statistics for Food and Agricultural Scientists

QA
276
P9

OTHER TITLES:

Statistical Quality Control for
Food and Agricultural Scientists
Subhash C. Puri,
Daniel Ennis,
Kenneth Mullen

Applied Statistics for Food and Agricultural Scientists

Subhash C. Puri
Canada Department of Agriculture
Ottawa, Ontario, Canada

Kenneth Mullen
University of Guelph
Guelph, Ontario, Canada

GKH

G. K. Hall Medical Publishers,
Boston, Massachusetts

WELSH AGRICULTURAL
COLLEGE
JOINT LIBRARY
20 JUL 1983
ABERYSTWYTH COLLEGE
OF FURTHER EDUCATION

Copyright © 1980 by G. K. Hall & Co.

G. K. Hall Medical Publishers
70 Lincoln Street
Boston, Massachusetts 02111

All rights, including that of translation into other languages, reserved. Photomechanical reproduction (photocopy, microcopy) of this book or parts thereof without special permission of the publisher is prohibited.

80 81 82 / 4 3 2 1

Bibliography
Includes index.
1. Mathematical statistics. I. Mullen, Kenneth, joint author. II. Title.
QA276.P85 5'9.5 79-15151
ISBN 0-8161-2161-3

Dedicated to our teachers who taught us and encouraged us to think:

Dean W. F. Forbes, M. L. Tiku, K. R. Shah
Dean C. H. Cramer, Boyd Harshbarger, Arthur Klingel

CONTENTS

Preface

There is no dearth of good textbooks on statistics, both applied and mathematical. In fact, few subjects have experienced such a bibliographic explosion. Notwithstanding, we have ventured into writing another textbook on applied statistics which, we hope, has some unique features. First, we have attempted to cover many different topics under one cover. Many of our colleagues working in the field of food and agricultural research have spoken of the inconvenience of searching through several textbooks to locate essential statistical methods. We hope this text will alleviate this problem and provide the researcher with a basic set of most commonly used statistical techniques. Second, we have directed our attention to the needs of food and agricultural scientists for whom few applied statistics books are available. Last, we have endeavored to keep the book as applied as possible, avoiding mathematical jargon, formulas, or their derivations. It is our belief that a knowledge of high school mathematics should be sufficient for understanding and using the statistical methods in this book. The book is intended as a working manual. Moreover, we hope that the simplicity and readability, coupled with a coverage of a variety of topics, will extend the usefulness of the book beyond the confines of food and agricultural sciences. In this sense, the book may be regarded as a general applied statistics text.

College and university students and teachers, research scientists, and those engaged in food industries and related organizations will find this book useful. In fact, it should be of interest to anyone wishing to employ statistical techniques to analyze data.

Another novel feature is the sequence of topic presentation. It seems logical to start with scientists who have a set of data they wish to analyze. They are encouraged to use a given sequence for analyzing data. First, the data are organized and tabulated to make them presentable. Then the data are graphed in order to study their distribution pattern. Next, the location and dispersion parameters are calculated. It is desirable to understand the methods of analyzing the data. The validity of the relevant parameters and their applicability within defined limits are assessed. Interpretation of results and report writing follow as a final phase of data analysis.

We wish to thank all those who helped us, either directly or indirectly, in writing this text. First, we are indebted to those scholars in the past and present whose abilities and perseverance have made possible the development of the available techniques and ideas. The authors' special thanks are due to Dean W. F. Forbes of the Faculty of Mathematics, University of Waterloo, Ontario, Canada, and Dr. A. R. Sen of Canadian Wildlife Service, Canadian Department of Fisheries and Environment, whose encouragement, comments, and suggestions had a great influence on the writing of the book. We are particularly thankful to Dr. Steve Brown of the University of Waterloo, A. K. Sahney of the Ministry of Agriculture, Government of Jamaica, H. S. Bajaj of Canadian Department of Indian and Northern Affairs, and B. S. Nagpal of Canadian Department of Fisheries and Environment, for their helpful suggestions and comments. We are also pleased to acknowledge the contributions of the editorial staff of G. K. Hall & Company. The authors alone, of course, are responsible for any errors or omissions. Last, but not the least, we thank Leah Holmes and Linda Hohner for their careful and patient typing of the manuscript, and our families for their patience, encouragement, and endurance.

S. C. Puri
K. Mullen

Applied Statistics for Food and Agricultural Scientists

1.

Introduction

Statistics deals with the development and application of methods and techniques for the collection, presentation, analysis, and interpretation of data. We collect data to obtain quantitative information about past results or to present observed phenomena and make predictions about their future behavior. The usefulness of statistics lies in its applicability to almost all fields of scientific study. Statistics can be broadly classified into two areas: theoretical statistics and applied statistics. Theoretical statistics provides a mathematical basis, whereas applied statistics deals with the applications of statistical methods for analyzing data.

In any scientific investigation, a food or agricultural scientist is generally faced with such problems as how to design an experiment efficiently to obtain valid data, how to analyze data properly through statistical methods, how to make relevant interpretations, and, finally, how to gauge the reliability of the results. The question of reliability arises because the investigator is generally dealing with a single sample of results and often wonders if repeated experiments would yield consistent results. Would all similar experiments provide reasonably consistent answers? This book offers guidelines for rational and orderly solutions to these problems. In this chapter we give some basic definitions and concepts, while the following chapters are devoted to applied statistical techniques essential for analyzing data.

1.1 Population and Sample

A *population* (sometimes called a *universe*) encompasses all possible measurements of the particular characteristic under consideration. A population can be finite or infinite, real or hypothetical. We may speak of the population of human beings in Canada, the height of men of a certain age group in a locality, the number of apples in an orchard, the number of grains of sand in the world, and so on.

A population is described by its *parameters.* If the population being studied is the height of all Canadian males over 21 years of age, then some parameters are the average height of the population, the range of heights of the population, or the most frequently occurring height of the population.

A *sample* is a subset of the population. It is made up of a selected number of individuals, each of which is a member of the population. Clearly, a sample will not tell us everything about the parent population from which it has been drawn, but an attempt is made to obtain information, as precisely as possible, about the parameters of the population. To do this requires good sampling procedure. An important type of sample in statistics is a *random sample.* A random sample is one that has been selected by a random process and is such that each measurement in the population has an equal and independent chance of being included in the sample. The theory of sampling is developed in a later section.

Whereas a population is described by its parameters, a sample is described by its *statistics.* If, from the population of heights of Canadian males over 21 years old, a sample of 1000 heights is obtained, some possible statistics are the average of the sample, the range of the sample, or the most frequently occurring value in the sample.

1.2 Continuous and Discontinuous Data

A measurable characteristic that changes from one object or individual to another is called a *variable.* Individual measurements of a variable are called *variates.* If the variable can assume only one value, it is called a *constant.*

In general, variables are of two kinds: *continuous* and *discontinuous,* or *discrete.* A discrete variable can assume only specific values (usually integers) and involves counting. For example, the number of cows on a farm or the number of peas in a pod are discrete variables. A continuous variable can take any value within a given range and involves measurements. Temperature readings and seasonal rainfall are continuous variables.

1.3 Scales of Measurement

Quantification of data must be done in accordance with some well-defined system. There are four commonly used scales of measurements: nominal, ordinal, interval, and ratio. The scale of measurement also influences the method applied for analyzing the data. A more detailed discussion of the subject is given in the chapter on Nonparametric Statistics.

1.3.1 Nominal

A variable on a nominal scale is classified by numerical names or symbols, such as high or low, dead or alive, left-handed or right-handed, Christian or Muslim. Even though a measurement on a nominal scale is the only measurement possible in some cases, it is weaker than the measurements on other scales. It is much more informative to describe a person as 2 meters tall than to simply say that the person is tall.

1.3.2 Ordinal or Ranking

An ordinal scale is a series of ranks. Numbers on an ordinal scale carry more information than numbers on a nominal scale because they imply relative magnitude. The ordinal scale is used whenever an experimenter indicates greater than, less than, or uses a rating system such as 0, 1, 2, or 3. For example, in a tasting panel experiment for determining the juiciness of cooked turkey, the ordinal scale can be set as very juicy, moderately juicy, juicy, dry, or very dry.

1.3.3 Interval

When a scale has all the characteristics of an ordinal scale and the exact distance between any two numbers is known, the measurements are on an interval scale. Even when the interval scale is a relative measure, it is much more precise than the ordinal scale. In an interval scale the zero point and the units of measurement are arbitrary and the ratios have no meaning. The intervals between the numbers represent differences, not absolute quantities. Suppose we wanted to measure the physical strength of a group of men of certain age groups. Since we can assume that there is no such thing as zero physical strength, we can arbitrarily designate a zero point and then measure everyone relative to that arbitrary zero.

1.3.4 Ratio

A variable on a ratio scale has all the characteristics of an interval scale and, in addition, has a true zero point as its origin. The ratio of any two scale points is independent of the units of measurement. For example, if we weigh things in grams with a true zero point, then the ratio of two weights is independent of the units of measurement. The ratio scale is stronger than any other scale.

1.4 Summation Notation

In statistics it is frequently necessary to work with sums of large numbers, and it is desirable to have a simplified notation for summation. The Greek letter Σ is the general symbol for summation. If a variable is denoted by X, then the particular values of that variable are written as X_1, X_2, X_3, and so on. In general, if there are n observations, i.e., n values of X, the last is denoted by X_n. The sum of the n observations is written as

$$\sum_{i=1}^{n} X_i = X_1 + X_2 + \cdots + X_n$$

which reads "sum of the X-values from i equals 1 up to and including i equals n," or more simply, "summation of X sub i, i going from 1 to n."

The summation notation can easily be extended to the case of two subscripts. Suppose we have k treatments, $T_1, T_2, \ldots, T_k$ with n observations per treatment; then X_{ij} refers to the *j*th observed value of the *i*th treatment. The sum of all the observed values is represented by

$$\sum_{i=1}^{k} \sum_{j=1}^{n} X_{ij}.$$

The following example demonstrates some rules of summation.

Example 1.1 If $X_1 = 2, X_2 = 4, X_3 = 1, X_4 = 5$, find

(a) $\sum_{i=1}^{4} X_i$ (b) $\sum_{i=1}^{4} X_i^2$ (c) $\left(\sum_{i=1}^{4} X_i\right)^2$

(d) $\sum_{i=1}^{4} (X_i - 1)$ (e) $\sum_{i=1}^{4} (X_i - 1)^2$ (f) $\sum_{i=1}^{4} X_i^2 - 1$

Solution

(a) $\sum_{i=1}^{4} X_i = X_1 + X_2 + X_3 + X_4$

$= 2 + 4 + 1 + 5 = 12$

(b) $\sum_{i=1}^{4} X_i^2 = X_1^2 + X_2^2 + X_3^2 + X_4^2$

$= (2)^2 + (4)^2 + (1)^2 + (5)^2 = 46$

(c) $\left(\sum_{i=1}^{4} X_i\right)^2 = (X_1 + X_2 + X_3 + X_4)^2$

$= (2 + 4 + 1 + 5)^2 = 144$

(d) $\sum_{i=1}^{4} (X_i - 1) = (X_1 - 1) + (X_2 - 1) + (X_3 - 1) + (X_4 - 1)$

$= (2 - 1) + (4 - 1) + (1 - 1) + (5 - 1) = 8$

(e) $\sum_{i=1}^{4} (X_i - 1)^2 = (X_1 - 1)^2 + (X_2 - 1)^2 + (X_3 - 1)^2 + (X_4 - 1)^2$

$= (2 - 1)^2 + (4 - 1)^2 + (1 - 1)^2 + (5 - 1)^2 = 26$

(f) $\sum_{i=1}^{4} X_i^2 - 1 = (X_1^2 + X_2^2 + X_3^2 + X_4^2) - 1$

$= (2^2 + 4^2 + 1^2 + 5^2) - 1 = 45$

Example 1.2 If a is a constant, then

$$\sum_{i=1}^{n} a = na$$

Example 1.3

(a) $\sum_{i=1}^{N} (aX_i - b) = a \sum_{i=1}^{N} X_i - Nb$

(b) $\sum_{i=1}^{N} aX_iY_i = a \sum_{i=1}^{N} X_iY_i$

(c) $\sum_{i=1}^{N} (X_i - Y_i)^2 = \sum_{i=1}^{N} X_i^2 - 2 \sum_{i=1}^{N} X_iY_i + \sum_{i=1}^{N} Y_i^2$

(d) $\sum_{i=1}^{N} (X_i - \overline{X})^2 = \sum_{i=1}^{N} X_i^2 - N\overline{X}^2$

$$= \sum_{i=1}^{N} X_i^2 - \frac{\left(\sum_{i=1}^{N} X_i\right)^2}{N}, \text{ where } \overline{X} = \frac{\sum_{i=1}^{N} X_i}{N}$$

Exercises

1. If $X_1 = 2$, $X_2 = -1$, $X_3 = 3$, $X_4 = 1$, find

(a) $\sum_{i=1}^{4} (X_i^2 - X_i)$ (b) $\sqrt{\sum_{i=2}^{4} X_i^2 + 5}$ (c) $\sum_{i=1}^{4} (X_i^2 - 3X_i + 30)$

(d) $\sum_{i=1}^{4} (X_i - \overline{X})^2$, where $\overline{X} = \frac{1}{4}\left(\sum_{i=1}^{4} X_i\right)$

2. Find (a) $\sum_{i=1}^{N} 1$ (b) $\sum_{i=1}^{4} 4$

3. Write each of the following in summation form.
   (a) $X_1^2 + X_2^2 + X_3^2 - 1$
   (b) $(X_1^2 - 1) + (X_2^2 - 1) + (X_3^2 - 1)$
   (c) $2X_3 + 2X_4 + 2X_5$

4. If $X_1 = 2, X_2 = 3, X_3 = -1, X_4 = 4, X_5 = 7$, find

   (a) $\sqrt{\sum_{i=1}^{5} (X_i + 1)^2}$ (b) $\left[\sum_{i=1}^{5} (X_i - 1)\right]^2$

5. If $X_1 = 3, X_2 = 0, X_3 = 1, X_4 = 4$, find

   (a) $\sum_{i=1}^{4} (3X_i^2 - 1)$ (b) $\sum_{i=1}^{4} 4X_i^2 - 2$

   (c) $\sum_{i=1}^{4} \sqrt{X_i + 2}$ (d) $\sum_{i=1}^{4} (X_i^2 - X_i + 1)$

6. Write out in full the sum represented by each of the following expressions.

   (a) $\sum_{i=1}^{3} 3X_i (X_i + 3)$

   (b) $\sum_{i=2}^{5} X_i^2 - 2$

   (c) $\sum_{i=1}^{4} f_i (X_i - 4)$

2.

Organization and Presentation of Data

Statistical data from a sample usually consist of a very large number of observations. To obtain any meaningful information, this raw, unorganized set of values has to be summarized concisely. This is achieved through tabulating the data in a *frequency table* and presenting them graphically. The graphical representation provides the distribution pattern of the data and helps determine the method to be used in analyzing them.

2.1 Frequency Table

The word "frequency" means the number of times that a particular observation is repeated. Data are summarized by grouping the variates into a number of classes and determining the *class frequency,* the number of individuals belonging to each class. Such a tabular arrangement of data is called a *frequency distribution* or *frequency table.* To illustrate the method of setting up a frequency table, consider the data in table 2.1 which give the percent crop loss due to drought as reported by 60 farmers in a particular area.

Example 2.1 From the data of table 2.1, construct a frequency distribution. To construct a frequency distribution from these data,

Table 2.1 *Percentage of crop loss for 60 farmers.*

17.1	6.7	9.1	10.7	13.1	10.8	10.8	9.3	12.3	6.2	14.2
10.0	14.4	10.6	10.9	8.1	10.1	10.1	11.0	13.6	11.1	7.3
14.6	11.2	11.3	4.6	11.6	12.8	12.8	11.8	16.1	12.1	9.9
6.9	12.3	14.3	12.0	9.6	12.2	12.2	13.3	12.4	16.9	8.9
12.6	11.9	11.6	14.1	10.0	12.0	12.0	13.9	12.4	10.4	12.9
14.9	10.2	8.7	18.4	13.7	13.7	13.7	15.1	12.6	7.9	16.4

we must first decide on two things: the desired number of classes or intervals, and the size of each interval. The number of intervals used depends largely on the nature of the data to be summarized. It is generally recommended to use between 5 and 15 intervals. A convenient approximation is provided by the empirical rule given by Sturges, according to which the number of intervals $= 1 + 3.3 \log_{10} N$, where N is the total number of observations. Applying the rule to the data in table 2.1, we find that the number of intervals $= 1 + 3.3 \log_{10} 60 = 1 + 3.3 \times 1.7782 = 6.87 \simeq 7$.

The next thing to decide is the size of the interval, or the class width. For convenience and simplicity, all classes should have the same class width, which can be approximated as follows:

$$\text{class width} = \frac{\text{range}}{\text{number of intervals}}$$

where range is equal to the largest observation minus the smallest observation. For our crop loss data, the highest value is 18.4%, the lowest value is 4.6%, and the range is 13.8%. Therefore, the approximate class width is 1.97 which, for convenience, may be rounded off to 2.00. Since the first class should contain the smallest observation, it must begin with 4.6 or less. A frequency table for the data in table 2.1 is shown in table 2.2.

The midpoint between the upper and lower boundaries of a class is called the *class midpoint* and is shown in column 2 of table 2.2. In columns 4 and 5 of table 2.2 are given the *class frequency* and *cumulative frequency,* respectively. The class frequency is the number of observations falling within a given class interval. The cumulative frequency for a given class is the sum of frequencies of all classes up to and including the given class.

Table 2.2 *Frequency distribution for percentage of crop loss.*

Class boundaries (%)	Class midpoint (X)	Tally	Frequency (f)	Cumulative frequency
4.5– 6.5	5.5	11	2	2
6.5– 8.5	7.5	~~1111~~	5	7
8.5–10.5	9.5	~~1111~~ ~~1111~~	10	17
10.5–12.5	11.5	~~1111~~ ~~1111~~ ~~1111~~ ~~1111~~	20	37
12.5–14.5	13.5	~~1111~~ ~~1111~~ 1111	14	51
14.5–16.5	15.5	~~1111~~ 1	6	57
16.5–18.5	17.5	111	3	60

NOTE: See Table 2.1.

2.2 Graphical Representation of Data

Once the data have been tabulated in a frequency table, the next step is to represent them graphically. Several types of graphical representations are used as aids in the study of frequency distributions. A few commonly used ones are presented here.

2.2.1 Histogram

A histogram is constructed by erecting rectangles on the class intervals marked off along the horizontal axis. The areas of the rectangles are proportional to the class frequencies. For equal class intervals, the vertical scale represents frequency. The histogram for the data in table 2.2 is shown in figure 2.1.

2.2.2 Frequency Polygon

In a frequency polygon, the value of a variate is concentrated at the midpoint of the class interval. The class midpoints are marked along the x-axis (horizontal axis) and the ordinates (vertical axis) are taken proportional to the frequencies. These points are joined by straight lines, and a frequency polygon is obtained. The frequency polygon for the data in table 2.2 is shown in figure 2.2

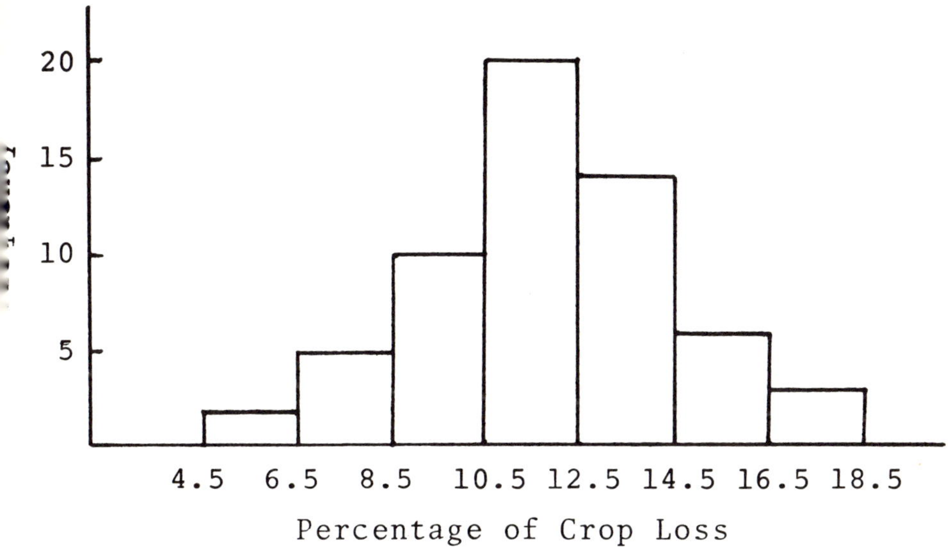

Fig. 2.1 *Histogram for the frequency distribution of percentages of crop loss of 60 farmers (see Table 2.2).*

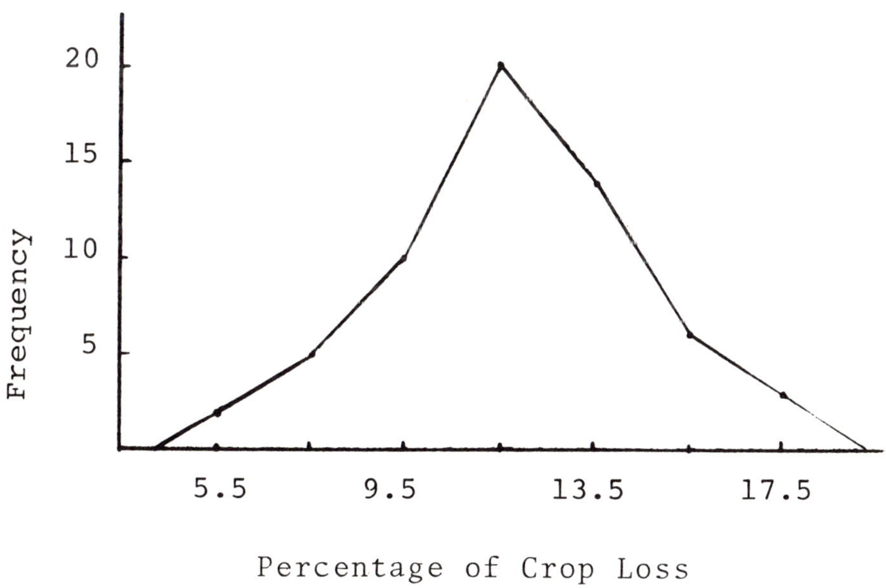

Fig. 2.2 *Frequency polygon for the data of Table 2.2.*

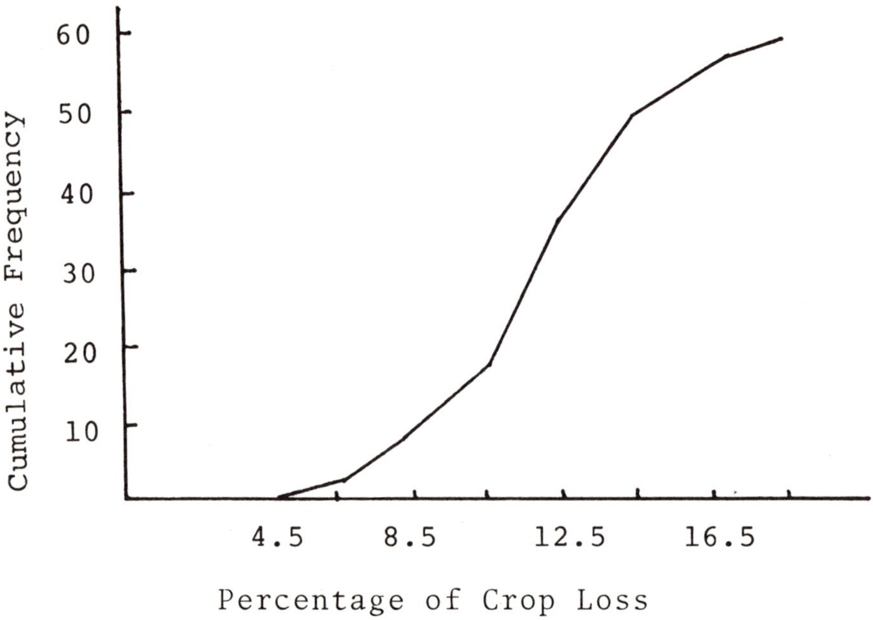

Fig. 2.3. *Cumulative frequency polygon (ogive) for the data of Table 2.2.*

2.2.3 Cumulative Frequency Polygon or Ogive

We may be more interested in the frequency of observations up to and including a given value rather than simply the frequency at that value. In that case, in place of frequency we deal with the cumulative frequency. The curve so obtained is called a *cumulative frequency polygon* or *ogive*. The ogive for the data in table 2.2 is shown in figure 2.3.

2.2.4 Frequency Curve

If the number of observations in the sample is large, and the class intervals can be made small, the histogram or the frequency polygon becomes a smoother curve. At the limit (for large populations), the sample and population are one and the same, and the histogram or frequency polygon becomes a smooth curve called a *frequency curve,* which shows the shape of the population. Some important types of frequency curves that occur in practice are given below.

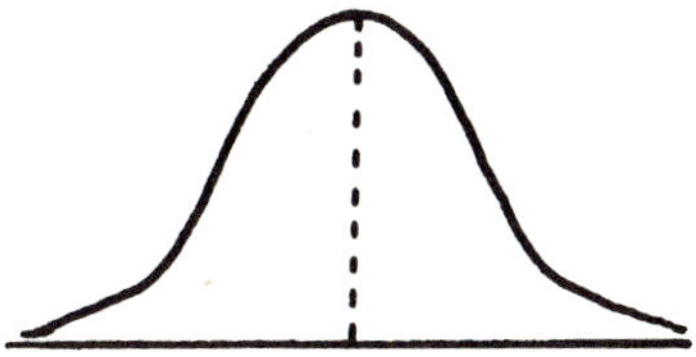

Fig. 2.4 *Symmetrical unimodal curve.*

Symmetrical Unimodal Curve
This curve has a maximum point in the middle and is symmetrical on both sides of the ordinate at this point. (A *bimodal curve* has two maximum points.)

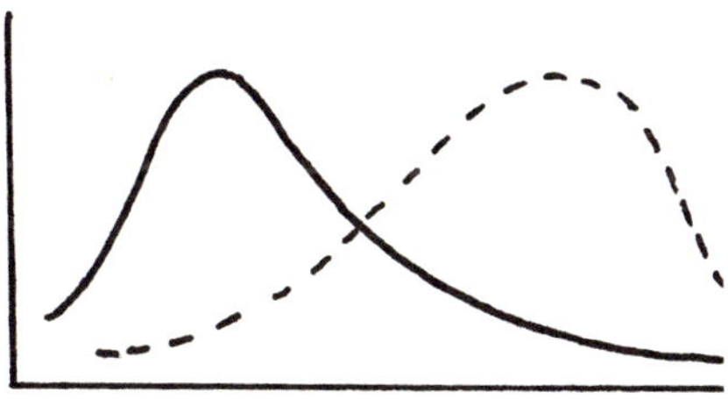

Fig. 2.5 *Skew curves.*

Moderately Asymmetrical (Skew) Distribution
In this case the frequency falls more rapidly on one side of the maximum than on the other.

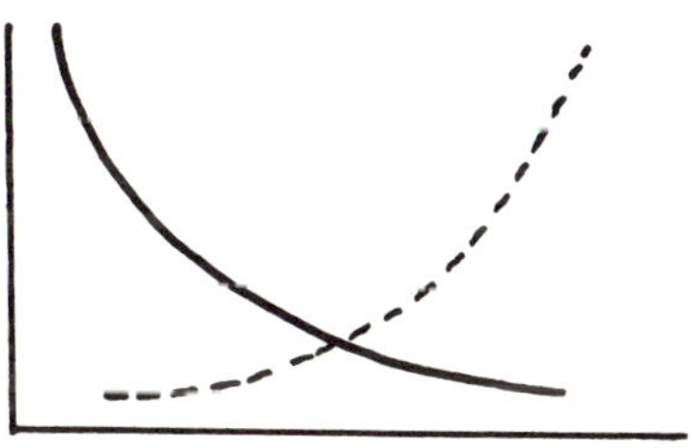

Fig. 2.6 *Extremely skew curves.*

Extremely Skew or J-Shaped Distribution
In these distributions the maximum frequency is at one end of the distribution. They do not exhibit symmetry about any ordinate.

U-Shaped or Antimodal Curves
These distributions have a minimum frequency with rising frequencies on either side of the minimum. They may or may not be symmetrical about this point. Such distributions occur, for instance,

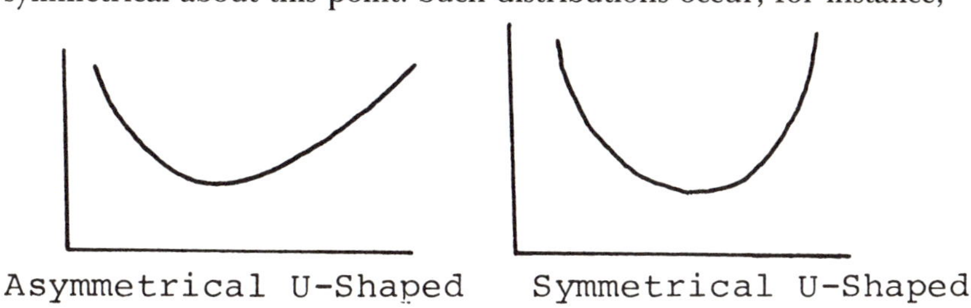

Fig. 2.7 *U-Shaped curves.*

when hospital patients are arranged according to age, since children and old people are more likely to be ill than other age groups.

2.2.5 Bar Graph

A bar graph is constructed by erecting bars of equal width either horizontally or vertically. Bar diagrams can be used in cases of discrete series, series of individual observations, spatial, and condition series. They cannot be used in continuous series. Bars are not suitable for representing the long-period time series. Whenever there is continuity in data, bar diagrams should not be used. This is the reason why bars are kept separate from each other. If they are joined together, the diagram would indicate data continuity and also become distorted and misleading. Vertical bar graphs are more commonly used than horizontal bar graphs. They represent the magnitude of data in relation to time. Two other variations in the horizontal or vertical bar graph are the component bar graph and the grouped bar graph.

Vertical Bar Graph
Table 2.3 gives the number of cans of peas found mutilated in a day's production in a cannery over a period of six days. Figure 2.8 is a bar graph for the same data.

Table 2.3 *Number of cans found mutilated during inspection.*

Days	Number of cans mutilated
1	50
2	62
3	85
4	59
5	112
6	150

Component and Grouped Bar Graphs
Table 2.4 gives the approximate figures for Canadian apple exports and imports for the 6-year period 1964–69.

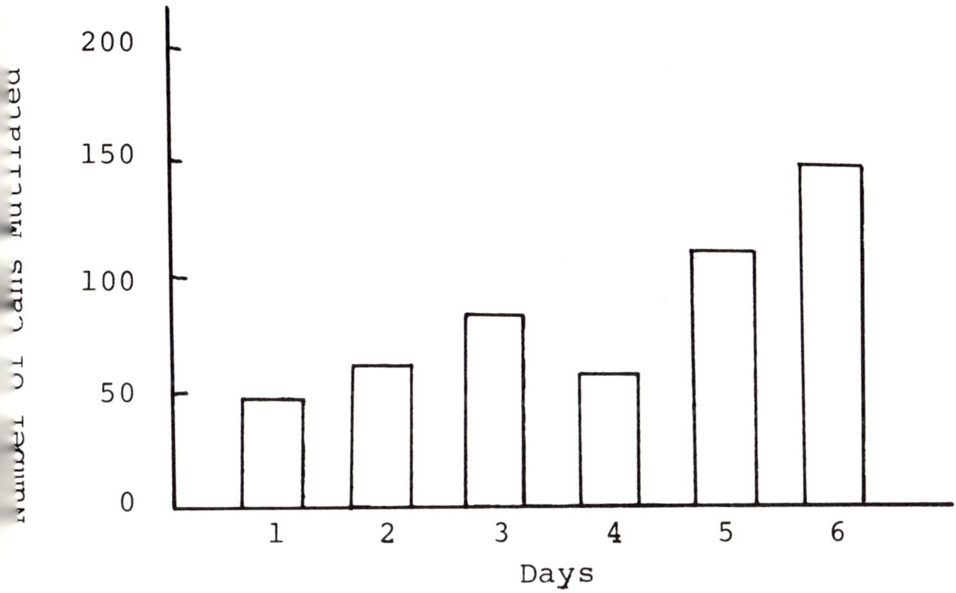

Fig. 2.8 *Bar graph for the data in Table 2.3.*

Table 2.4 *Exports and imports of apples.*

Year	Exports	Imports
	(millions of pounds)	
1964	155	35
1965	130	50
1966	112	60
1967	140	45
1968	170	70
1969	135	55

Figures 2.9 and 2.10 show, respectively, the component bar graph and grouped bar graph for the data in table 2.4.

2.2.6 Pie Chart

The pie chart is a circle whose area is divided into segments. It is used to portray component relations. The various sectors of the circle represent component parts of an aggregate. A pie graph

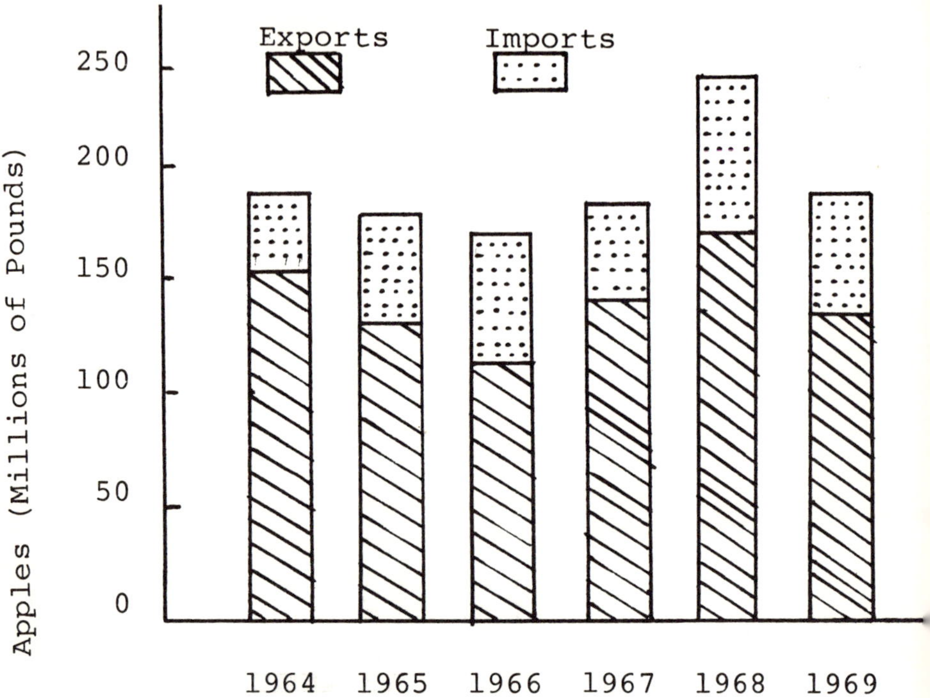

Fig. 2.9 *Component bar graph for the data in Table 2.4.*

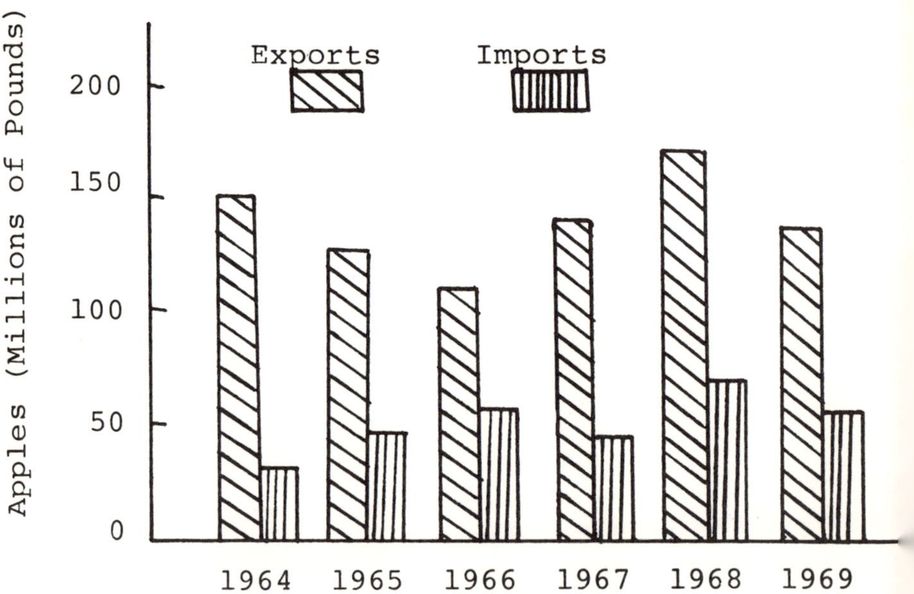

Fig. 2.10 *Grouped bar graph for the data in Table 2.4.*

should not consist of too many segments (between 5 and 7 is suggested). Each segment should be labeled clearly and concisely.

Table 2.5 shows the breakdown of major expenditures of an average family of two who own a single-family home. The pie graph is shown in figure 2.11.

Table 2.5 *Breakdown of family expenditures.*

Expenditure	Percentage
Home:	45%
Food:	21%
Clothing:	10%
Entertainment:	6%
Other:	18%

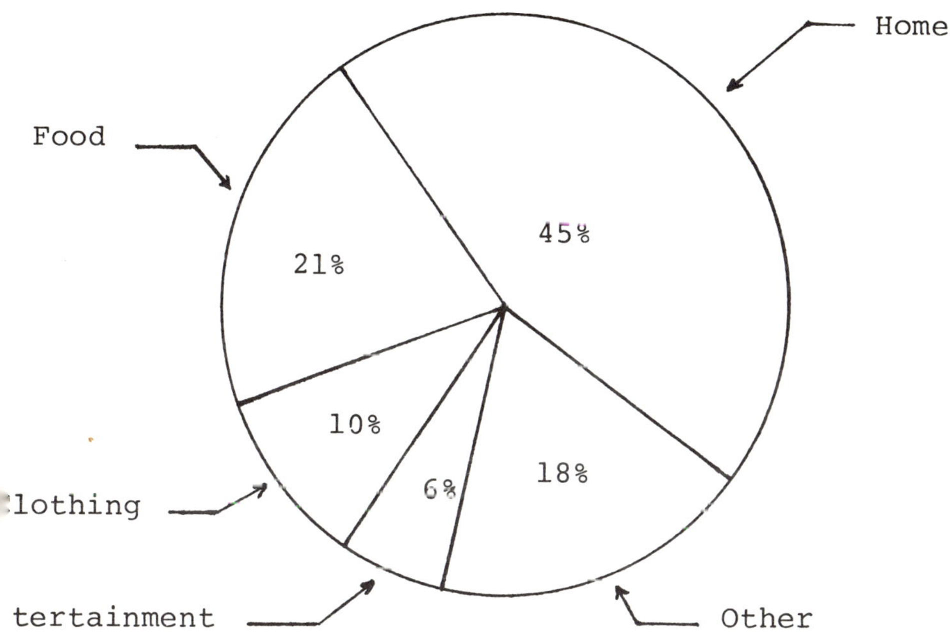

Fig. 2.11 *Pie chart for the data in Table 2.5.*

2.3 Cheating by Charting

Misleading graphical representation of data is not uncommon. Distortions can occur either unintentionally or be cleverly planned to stress a point. The most common distortion is due to a change in scale.

Example 2.2 The average percentage of defective eggs found during inspection over a period of one month at two different grading stations was 6.9% and 7.5%.

The bar graph in figure 2.12.1 shows the right way to present the data. However, if we wish to exaggerate the difference between the two defective rates, it can be done by using the broken vertical scale starting at 6.5 rather than at 0 as shown in figure 2.12.2.

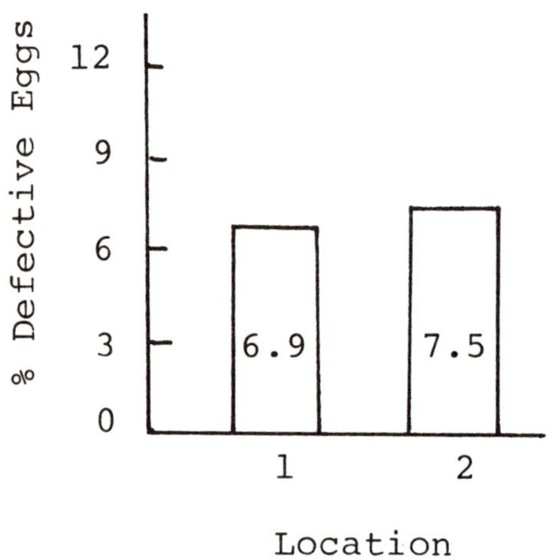

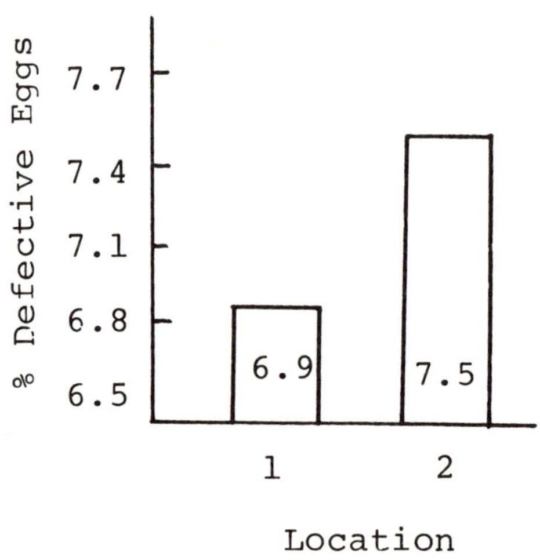

Fig. 2.12.1, 2.12.2 *Bar graph showing percentage of defective eggs.*

Exercises

1. The percentage of germination of oat seeds, as found on germination trial by five different laboratories, is as follows:

	Laboratories				
	1	2	3	4	5
Percent germination	89	83	85	89	90

Draw a vertical bar graph for the data.

2. The following data give the net weight in ounces for a sample of 60 cans of baby food produced in a processing plant.

9.8	10.4	10.0	9.6	10.1	10.3
9.9	9.7	10.1	10.0	10.0	10.0
10.0	9.8	9.7	10.1	10.2	10.0
9.7	10.0	10.1	9.9	10.2	10.0
10.1	9.9	10.2	10.3	9.8	10.4
10.0	10.1	9.9	10.3	10.0	10.1
10.2	9.8	9.9	10.0	10.3	10.1
9.8	10.0	10.1	10.2	10.0	10.0
9.9	9.9	10.0	10.1	10.2	10.1
10.0	9.9	10.1	10.2	9.9	10.0

(a) Construct a frequency table for these data.
For the above frequency table, draw a
(b) line graph
(c) histogram
(d) frequency polygon
(e) cumulative frequency polygon

3. The following data give the percentage of defective apples found by an inspector during inspection of six lots of a shipment.

Lot	Percentage defective
1	3.4
2	3.6
3	4.1
4	3.9
5	4.2
6	3.1

Construct a frequency polygon and frequency curve for these data.

4. When commercial butter is graded, its quality is usually expressed by a scoring system wherein each characteristic has a potential value out of a total of 100 points, as follows:

Characteristic	Value
Flavor	45
Texture	15
Moisture	10
Color	10
Salt	10
Finish	10
Total	100

Construct a pie chart to represent the scoring system.

5. The following data, from the Livestock Market Review, give the number of cattle and calves slaughtered in inspected packing plants in Canada for selected years.

Year	Cattle	Calves
	(thousands)	
1966	2705	766
1967	2642	739
1968	2784	668
1969	2718	580
1970	2701	499
1971	2787	464

Construct a grouped bar graph for these data.

6. Construct a histogram for the frequency distribution of corn yield data of exercise 3, chapter 3.

7. The average price for bread for the 5-year period 1973–77 is as follows:

Year	1973	1974	1975	1976	1977
Price($)	0.26	0.32	0.39	0.45	0.45

Draw a bar graph for these data.

3.

Statistical Measures

Although graphical representation of data gives a general idea of its distribution, to compare similar distributions we must describe them arithmetically. Certain features give a general idea of the distribution and can be determined arithmetically. They include measures of location, or central tendency, and measures of dispersion, or variability.

3.1 Measures of Central Tendency

The most common characteristic of experimental data is the tendency to build up observations in the center of the distribution. This characteristic, known as central tendency, is usually expressed through measures of averages, the most common being the mean, the median, the mode, the geometric mean, and the harmonic mean.

3.1.1 The Mean

The arithmetic mean is the most commonly used measure of central tendency, and represents the center of a frequency distribution in much the same way that the center of gravity represents a balance point for the weight distribution in an object.

For a sample of observations $X_1, X_2, \ldots, X_n$, the *sample arithemetic mean* (or *sample mean* or, simply, the *mean*), denoted by $\overline{X}$ is

$$\overline{X} = \frac{X_1 + X_2 + \cdots + X_n}{n} = \frac{\sum_{i=1}^{n} X_i}{n}$$

Note: The number of variates in a sample is represented by n and the number in the population by N. Sample statistics, such as the sample mean, are usually represented by Latin letters and population parameters are usually represented by Greek letters.

For a population of N values, $X_1, X_2, \ldots, X_N$, the population mean, denoted by μ, is

$$\mu = \frac{X_1 + X_2 + \cdots + X_N}{N} = \frac{\sum_{i=1}^{N} X_i}{N}$$

Example 3.1 Production of condensed whole milk in Canada for the 5-year period 1968–72 was, respectively, (millions of pounds)

26, 22, 16, 16, 17

The arithmetic mean of yearly production is

$$\overline{X} = \frac{26 + 22 + 16 + 16 + 17}{5} = 19.4 \text{ million pounds}$$

Note: An interesting property of the arithmetic mean is that the sum of the deviations of the individual values from their mean is always equal to zero, that is,

$$\sum_{i=1}^{n} (X_i - \overline{X}) = 0$$

Thus, if we subtract the mean of 19.4 from each of the individual values, the resulting five deviations will add up to zero.

$$\sum_{i=1}^{5} (X_i - \overline{X}) = [(26 - 19.4) + (22 - 19.4) + (16 - 19.4) + (16 - 19.4) + (17 - 19.4)]$$
$$= 0$$

For grouped data where the variables $X_1, X_2, \ldots, X_n$ occur with respective frequencies $f_1, f_2, \ldots, f_n$, the mean is defined as

$$\overline{X} = \frac{X_1f_1 + X_2f_2 + \cdots + X_nf_n}{f_1 + f_2 + \cdots + f_n} = \frac{\sum_{i=1}^{n} X_if_i}{\sum_{i=1}^{n} f_i}$$

Example 3.2 The mean of the data in table 2.2, which represents percentage of crop loss due to drought, is

$$\overline{X} = \frac{(5.5)(2) + (7.5)(5) + (9.5)(10) + (11.5)(20) + (13.5)(14) + (15.5)(6) + (17.5)(3)}{2 + 5 + 10 + 20 + 14 + 6 + 3}$$
$$= 11.8\%$$

Weighted Arithmetic Mean

If the variables $X_1, X_2, \ldots, X_n$ are associated with certain weighting factors or, weights, $W_1, W_2, \ldots, W_n$, depending on some importance attached to the numbers, the weighted arithmetic mean is given by

$$\overline{X} = \frac{X_1W_1 + X_2W_2 + \ldots + X_nW_n}{W_1 + W_2 + \ldots + W_n} = \frac{\sum_{i=1}^{n} X_iW_i}{\sum_{i=1}^{n} W_i}$$

Example 3.3 In a beef cattle show, the judge assigns weights of 1, 2, and 3, respectively, to general appearance, weight, and health of the animals. A farmer's entry obtains the following scores:

General appearance	=	72
Weight	=	79
Health	=	83

Calculate the average score the farmer receives for the cattle.

$$\overline{X} = \frac{(1)(72) + (2)(79) + (3)(83)}{1 + 2 + 3} = 79.83 \simeq 80$$

3.1.2 The Median

The *median* is the halfway point in a distribution. For a set of n numbers $X_1, X_2, \ldots, X_n$ arranged in ascending or descending order of magnitude, the median, denoted by M, is the middle number of the set if n is odd, and the mean of the two middle numbers if n is even,

$$\text{Median} = M = X_{(n+1)/2}, \text{ if n is odd}$$

$$= \frac{X_{(n/2)} + X_{(n/2)+1}}{2}, \text{ if n is even.}$$

Note: We shall use $X_{(i)}$ to represent the *i*th smallest observation after the sample has been arranged from smallest to largest.

Example 3.4 The median of 11, 12, 14, 18, 19, 20, 26, 27, is

$$M = \frac{X_{(4)} + X_{(5)}}{2} = \frac{18 + 19}{2} = 18.5$$

In the case of data grouped into a frequency distribution, for a set of numbers $X_1, X_2, \ldots, X_n$ with respective frequencies $f_1, f_2, \ldots, f_n$, the median is obtained by simple interpolation using the formula

$$M = L + C\left[\frac{\frac{N}{2} - f_b}{f_m}\right]$$

where L = real lower limit of interval containing the median class

C = class width

N = total frequency $\sum_{i=1}^{n} f_i$

f_b = cumulative frequency below the interval containing the median

f_m = frequency of the median class, the median class being the lowest class for which the cumulative frequency exceeds $\frac{N}{2}$

Example 3.5 For the crop loss data in table 2.2, the median can be calculated as follows:

$$L = 10.5,\ C = 2,\ N = 60,\ f_b = 17,\ f_m = 20$$

$$M = 10.5 + 2\left[\frac{\frac{60}{2} - 17}{20}\right]$$

$$= 11.8\%$$

3.1.3 The Mode

The *mode* is the most frequently occurring observation in the data. In a frequency table, the modal class is the one with the highest frequency and is the highest point on a frequency curve. If a curve is symmetrical, the mode will be the midpoint of the modal class. However, if the curve is skewed, one class adjacent to the modal class will have larger frequency than the other adjacent class. The peak of the frequency curve will not be at the midpoint of the modal class, but will lie closer to the adjacent class with the larger frequency. Thus, we can locate the mode graphically as in figure 3.1.

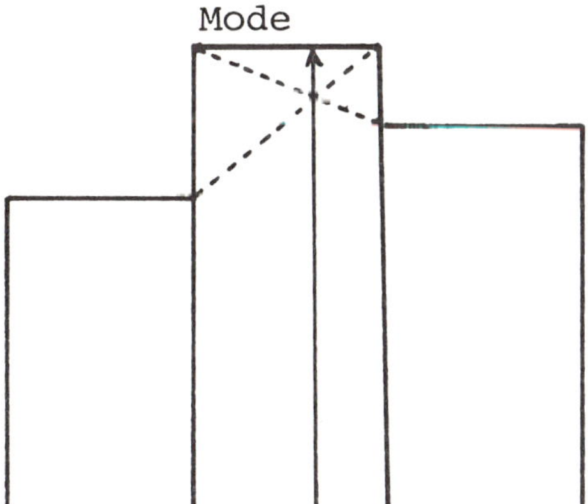

Fig. 3.1 *Graphic location of the mode.*

The formula for calculating the mode of grouped frequency data is given by

$$\text{Mode} = L + C\left[\frac{f_a}{f_a + f_b}\right]$$

where L = real lower limit of the modal class
C = class width
f_a = frequency of the class immediately above the modal class
f_b = frequency of the class immediately below the modal class

Example 3.6 Calculate the mode for the crop loss data in table 2.2.

$$L = 10.5,\ C = 2,\ f_a = 14,\ f_b = 10$$

$$\text{Mode} = 10.5 + 2\left[\frac{14}{14 + 10}\right] = 11.67\%$$

3.1.4 The Geometric Mean

The *geometric mean,* denoted by G, of a series of n observations X_1, $X_2, \ldots, X_n$ is given by

$$G = (X_1 \cdot X_2 \ldots X_n)^{\frac{1}{n}}$$

It is convenient to express this in logarithmic form; thus

$$\log G = \frac{1}{n}\sum_{i=1}^{n} \log X_i$$

For a frequency distribution, $G = (X_1^{f_1} \cdot X^{f_2} \ldots X_n^{f_n})^{\frac{1}{N}}$, where $N = \sum_{i=1}^{n} f_i$, or in logarithmic form, $\log G = \frac{1}{N}\sum_{i=1}^{n} f_i \log X_i$. The geometric mean is used mainly in averaging ratios or rates of change over time.

3.1.5 The Harmonic Mean

For a set of n numbers $X_1, X_2, \ldots, X_n$, the *harmonic mean,* denoted by H, is given by

$$H = \frac{n}{\sum_{i=1}^{n} \frac{1}{X_i}}$$

The harmonic mean is used in problems of averaging rates per unit of time.

3.2 Advantages and Disadvantages of Different Averages

Of all the different measures of central tendency, the most valuable and commonly used measure is the arithmetic mean. It is well-defined, based on all observations, and lends itself easily to algebraic treatment. It gives weight to all items in direct proportion to their sizes and, in most cases, is minimally affected by fluctuations of sampling. However, the arithmetic mean tends to give undue weight to extreme values and sometimes provides values that may not be possible, such as an average number of births per animal of 2.3 per year.

Two other measures of location used frequently, though not as frequently as the arithmetic mean, are the median and the mode. Like the mean, they are also easily calculated, and are well-defined. Sometimes the median is preferred over the mean because it does not give undue weight to very high or very low values and, unlike the mean, it can be calculated in cases where the end intervals are open. However, the median and the mode do not lend themselves easily to algebraic treatment.

The geometric mean and harmonic mean have somewhat limited areas of usage. The geometric mean is suitable where ratios or indexes are given, and the harmonic mean is useful for finding, say, the average speed when the speed for different parts of the distances are given.

In general, the arithmetic mean is the most convenient and widely used form of average and should always be used unless there are strong reasons against it.

3.3 Measures of Dispersion

While averages give an idea of the central tendency of the given distribution, it is also necessary to know how the values of the variable are dispersed about the central value. The commonly used measures of dispersion are the range, the mean deviation, the variance and standard deviation, and coefficient of variation.

3.3.1 The Range

The *range* for a set of observations is the difference between the largest and smallest observations. For a frequency table, the range is the difference between the upper boundary of the highest class and the lower boundary of the lowest class. Its use is very much limited, since it does not take into account the central tendency or the form of the distribution.

Example 3.7 The range of the three sets of data
(a) 3, 6, 9, 12, 15, 18, 21
(b) 3, 3, 3, 3, 3, 3, 21
(c) 3, 12, 12, 12, 12, 12, 21

is the same, namely $21 - 3 = 18$, but intuitively we regard the three sets as being quite different.

3.3.2 The Mean Deviation or Average Deviation

Another measure that immediately suggests itself is the average of the deviations of a set of numbers from their mean. But from section 3.1 we note that the sum of the deviations of a set of variates from their mean is equal to zero. Therefore, we consider the average of the absolute values of the deviations and get a useful measure of dispersion called the *mean deviation* from the mean. The absolute value of a number C is denoted by $|C|$ and is defined as $|C| = C$ if $C \geq 0$ and $|C| = -C$ if $C < 0$. This means that C is always positive. For example, $|-3| = 3$, $|5| = 5$.

Thus, for a set of n numbers, $X_1, X_2, \ldots, X_n$, whose mean is $\overline{X}$, the mean deviation, denoted by M.D. is defined as

$$\text{M.D.} = \frac{\sum_{i=1}^{n} |X_i - \overline{X}|}{n}$$

If $X_1, X_2, \ldots, X_n$ occur with frequencies $f_1, f_2, \ldots, f_n$, respectively, the mean deviation is given by

$$\text{M.D.} = \frac{\sum_{i=1}^{n} |X_i - \overline{X}| f_i}{\sum_{i=1}^{n} f_i}$$

The mean deviation has two properties that make it a good index of variability:

1. As the dispersion of the values increases or decreases, M.D. will become larger or smaller proportionately.

2. M.D. is sensitive to the position of every value in a distribution, that is, any displacement of any value influences the size of M.D.

Example 3.8 The total scores given by a dairy inspector to five samples of butter from five different churns of a creamery are as follows:

72, 75, 82, 92, 79

Calculate the mean deviation for these data.

$$\overline{X} = \frac{72 + 75 + 82 + 92 + 79}{5} = 80$$

$$\text{M.D.} = \frac{|72 - 80| + |75 - 80| + |82 - 80| + |92 - 80| + |79 - 80|}{5}$$

$$= \frac{8 + 5 + 2 + 11 + 1}{5} = 5.4$$

The average amount by which the scores deviate from the mean of the distribution is 5.4 score units.

3.3.3 The Variance and the Standard Deviation

The mean deviation provides some idea of the amount of variability in the data, but one has to make the arbitrary decision to disregard the negative signs associated with those deviations less than the mean. To avoid this, another measure of dispersion, known as the *variance,* considers the sum of the squared deviations from the mean.

For a set of n observations, $X_1, X_2, \ldots, X_n$, the *sample variance,* denoted by S^2 is defined as

$$\text{variance} = S^2 = \frac{\sum_{i=1}^{n} (X_i - \overline{X})^2}{n - 1}$$

A more convenient way of writing the variance is as follows:

$$S^2 = \frac{\sum_{i=1}^{n} X_i^2 - \frac{\left(\sum_{i=1}^{n} X_i\right)^2}{n}}{n - 1}$$

For a population of values $X_1, X_2, \ldots, X_N$, the *population variance,* denoted by σ^2 (sigma squared) is

$$\sigma^2 = \frac{\sum_{i=1}^{N} (X_i - \mu)^2}{N}$$

Unlike the mean, which can be likened to a balance point on a seesaw, the variance has no such physical analogue. Sometimes the variance is associated with the moment of inertia—an index of the torque produced by a rotating body.

The square root of the variance is called the *standard deviation.* The *sample standard deviation* is

$$S = \sqrt{\frac{\sum_{i=1}^{n} (X_i - \overline{X})^2}{n - 1}}$$

If the observations are measured in grams or inches, then the variance, being the square of the deviations, is measured in gm^2 or in^2. The standard deviation is measured in the same units as the original data.

Example 3.9 The average wholesale price (cents per pound) of creamery butter in Canada for the five-year period 1967–71 was as follows:

59.0, 62.5, 63.5, 65.0, 66.1

Calculate the variance and the standard deviation.

Here, $\Sigma X_i^2 = (59.0)^2 + (62.5)^2 + (63.5)^2 + (65.0)^2 + (66.1)^2$

$= 20013.71$

$\Sigma X_i = 59.0 + 62.5 + 63.5 + 65.0 + 66.1 = 316.0$

$n = 5$

Therefore,

$$\text{Variance} = S^2 = \frac{20013.71 - \dfrac{(316.0)^2}{5}}{4} = 10.63¢$$

Standard deviation $= S = \sqrt{10.63} = 3.26¢$

In the case of a grouped frequency distribution, where a set of numbers $X_1, X_2, \ldots, X_n$ have their respective frequencies $f_1, f_2, \ldots, f_n$, the variance is given by

$$S^2 = \frac{\sum_{i=1}^{n} (X_i - \overline{X})^2 f_i}{\sum_{i=1}^{n} f_i - 1}$$

or

$$= \frac{\sum_{i=1}^{n} X_i^2 f_i - \dfrac{\left(\sum_{i=1}^{n} X_i f_i\right)^2}{\sum_{i=1}^{n} f_i}}{\sum_{i=1}^{n} f_i - 1}$$

Example 3.10 Calculate the variance for the crop loss data in table 2.2. Table 3.1 shows the details of the calculations for the variance.

Table 3.1 *Calculation of variance for the data in Table 2.2.*

Class midpoint (X_i)	Frequency (f_i)	$X_i f_i$	$X_i^2 f_i$
5.5	2	11.0	60.50
7.5	5	37.5	281.25
9.5	10	95.0	902.50
11.5	20	230.0	2645.00
13.5	14	189.0	2551.50
15.5	6	93.0	1441.50
17.5	3	52.5	918.75
	60	708.0	8801.00

$$S^2 = \frac{8801.00 - \dfrac{(708.0)^2}{60}}{60 - 1} = 7.57$$

$$S = \sqrt{7.57} = 2.75$$

3.3.4 The Coefficient of Variation

The *coefficient of variation,* denoted by C, is a measure of relative dispersion and is given by

$$C = \frac{S}{\bar{X}}$$

It is generally expressed as a percentage. The utility of this measure lies partly in the fact that the mean and standard deviation tend to change together in many experiments. A knowledge of relative variation is valuable in evaluating the experiments. Note that the coefficient of variation is independent of the units used because both the numerator and the denominator of the expression for C are dimensionally equivalent, so that the units of measurement cancel in the ratio.

The meaning, use, and interpretation of C depends largely on the nature of the experiment and the scientist's relative experience in handling data. The coefficient of variation is a good measure of relative dispersion, but relative dispersion is not always a good measure of variability. The validity of using C depends on the assumption that the absolute dispersion in biological distributions increases in direct proportion to the mean in variates of essentially the same variability. Lower values of C are indicative of inadequate samples, whereas large values of C usually indicate that the sample was mixed, i.e., it contained all sorts of different elements and characteristics.

Consider the crop loss data of table 2.1. The mean and standard deviation for these data from examples 3.2 and 3.10 are

$$\overline{X} = 11.8, S = 2.75$$

Thus,

$$C = \frac{S}{\overline{X}} = \frac{2.75}{11.8} = 0.23 \text{ or } 23.0\%$$

Exercises

1. The following data represent laboratory analysis by four different laboratories of the fat content of skim milk powder produced by five different companies.

Company	Laboratory			
	1	2	3	4
A	.601	.588	.755	.706
B	.735	.689	.938	.831
C	.606	.580	.806	.596
D	.670	.648	.812	.721
E	.805	.840	.962	.839

Find
(a) the mean and variance for each company
(b) the mean and variance for each laboratory
(c) the mean, standard deviation, and coefficient of variation for the entire data

2. From the frequency distribution obtained in exercise 2, chapter 2, calculate the (i) mean, (ii) median, (iii) mode, (iv) range, (v) variance, (vi) standard deviation, and (vii) coefficient of variation.

3. The following data represent the frequency distribution of the yield of 100 identically treated plots of corn (in lbs).

Class boundaries	No. of plots
7.5–10.5	8
10.5–13.5	10
13.5–16.5	15
16.5–19.5	20
19.5–22.5	25
22.5–25.5	16
25.5–28.5	6
	Total = 100

Find the (i) mean, (ii) median, (iii) mode, (iv) variance, and (v) standard deviation.

4. Calculate the mean, standard deviation, and the coefficient of variation for the following frequency distribution.

Class boundaries	Frequency
5– 9	2
9–13	4
13–17	8
17–21	9
21–25	6
25–29	1

5. Two fertilizers, A and B, were each given to four plots of tomato plants. They yielded the following results.

	Fertilizer	
Plot	A	B
1	35	30
2	39	29
3	37	28
4	38	31

Calculate the mean and variance for the result of each fertilizer.

6. The weights in grams of 10 apples picked out at random from a consignment are as follows:
156, 128, 93, 106, 82, 115, 100, 95, 126, 112
Calculate the mean, mean deviation, and standard deviation of the data.

7. The following data give the approximate weekly consumption per person of some selected articles of food, their quantity, and value. Calculate (i) the unweighted mean of prices in column 3, and (ii) the weighted mean obtained by weighing each price by the quantity consumed.

Articles of food (1)	Quantity consumed in pounds (2)	Price in dollars per pound (3)
Potatoes	4.00	0.28
Sugar	1.00	0.60
Coffee	0.60	4.50
Flour	3.50	0.21
Beef	5.00	3.10

8. A sampling experiment with 55 peas results in the following frequency distribution of the number of peas in each pea pod

5	3	4	4	4	3	4	5	5	4	2
4	3	4	5	6	1	4	3	6	4	2
5	4	4	3	7	4	1	2	3	4	5
3	6	5	3	6	4	3	6	4	5	3
4	4	4	3	2	5	6	2	4	3	4

Set up a frequency table for these data and draw a line graph. Calculate the mean, median, variance, and standard deviation for the distribution.

9. The ratios of the prices of butter, milk, eggs, potatoes, and bread in 1965 to those in 1977 are 0.87, 1.10, 0.75, 0.65, and 1.20, respectively. Calculate the geometric mean for the data.

10. The average prices per month of 1 lb of beef steak for a four-month period are (in dollars).

3.00, 2.85, 3.20, 3.50

A family spends an equal amount of money on beef steak for this four-month period. Calculate the harmonic mean for the data.

11. The following table gives the yield per plant of five tomato plants of a certain variety.

Plant number	Yield (gm)
1	1325
2	1412
3	1492
4	1375
5	1296

Calculate the mean and standard deviation for the data.

12. The number of poppy plants found in 10 fields of oats, each given a different treatment, was as follows: 85, 70, 15, 75, 500, 20, 45, 250, 40, 36. Calculate the arithmetic mean, geometric mean, and harmonic mean, and make a comparative study of the three measures.

13. Coding of data: Sometimes when data are expressed in large numbers and mechanical means for carrying out statistical calculations are not available, the tedium of calculating can be lessened by coding the data into a new scale. The types of coding most commonly used are additive and multiplicative coding.

Additive coding consists of adding or subtracting a constant, say c, from each observation to obtain a new set of numbers that are comparatively less tedious to handle. For example, consider $y_i = x_i \pm c$, $i = 1, 2, \ldots, n$, where x_i's are original measurements and y_i's are coded measurements, c being a constant. Then

$$\bar{y} = \bar{x} \pm c \quad \text{or} \quad \bar{x} = \bar{y} \pm c$$

The variance of the data remains unaffected by the additive coding.

Similarly, in multiplicative coding, the measurements are multiplied or divided by a constant. In this case, we have

$$y_i = cx_i,$$

$$\bar{y} = c\bar{x} \quad \text{or} \quad \bar{x} = \frac{1}{c}\bar{y}$$

$$\text{var}(y) = c^2 \text{ var } (x)$$

$$\text{i.e., } s_y^2 = c^2 s_x^2 \text{ or } s_x^2 = \frac{1}{c^2} s_y^2$$

Calculate the mean and variance of the following after coding the data.

(a) x_i: 478, 482, 471, 473, 479
Subtract c = 470 from each observation to code the data.

(b) x_i: 0.01, 0.03, 0.08, 0.02, 0.04
Multiply each of the x_i's by c = 100 to code the data.

14. Calculate the mean and variance for the following data using the coding method.
76, 79, 75, 82, 69, 75

4.

Probability

A central problem in statistics is to use information in a sample to gain knowledge about the population. For instance, what does the knowledge of $\overline{X}$ tell us about the value of μ, the population mean. These problems are discussed in more detail subsequently. Suffice it to say for now that all statements about the population are made with varying degrees of probability. For this reason, a knowledge of elementary probability is necessary to understand later statistical results.

Events that are common or likely to occur are said to have a high probability of occurring and those that are rare or unlikely are said to have a low probability of occurring. The general idea of probability is used in everyday speech when, for instance, one says "Rain is very likely today," (or probable) or "The chance of my plane to San Francisco crashing is negligibly small" (i.e., unlikely). Statisticians replace the vague words "very likely" and "negligibly small" with numbers from zero to one, so that low and high probabilities are represented by numbers which are near zero for low probabilities, near one for high probabilities, and intermediate for other probabilities.

4.1 Classical Probability

Since probability developed from the need of early gamblers to calculate their probabilities of winning and losing, it seems fitting to present the method of calculating probabilities using this gambling (or classical) approach. If a procedure gives rise to n equally likely outcomes of which r have attribute A, then the probability of A is r/n.

Example 4.1 If a consignment of 3000 lettuces is composed of 2/3 head and 1/3 leaf (so that there are 2000 head lettuces and 1000 leaf lettuces), and we randomly pick a single lettuce from the consignment, the probability that it will be a head lettuce is 2000/3000 or 2/3. Similarly, the probability that it will be a leaf lettuce is 1/3.

4.2 Probability as a Relative Frequency

The classical definition of probability is somewhat limited, since to calculate any probabilities we need to know (i) the value of n and (ii) that each outcome is equally likely. Since most food and agricultural problems do not satisfy these requirements, a more pragmatic definition of probability, called the *frequency concept* of probability, is used. If an event of interest, A, occurs r times in n trials, and if the ratio r/n approaches a limit as n becomes large, then r/n is called the probability of A, and we write $P(A) = r/n$.

One sees from either definition of probability that (i) $P(A) \geq 0$, (ii) $P(A) \leq 1$, and (iii) if $\overline{A}$ represents the complement of A, then $P(\overline{A}) = 1 - P(A)$. Throughout this text and, in fact, for most working definitions, we shall interpret probability as a *relative frequency* in a large number of trials. While this definition is not completely free from controversy because of a dispute concerning the word "limit," the student will find it useful, convenient, and practical and it will never lead one to the wrong conclusion. We state again: *When we talk of the probability of an event A, we mean the relative frequency of A in a large number of similar (or repeated) trials.* Two special kinds of events merit attention. If an event A is impossible, then $r = 0$ and $P(A) = 0$. If an event A is certain (i.e., it occurs at every trial), then $r = n$ and $P(A) = 1$.

Example 4.2 Cucumbers are to be graded into one of three classifications. Class I (acceptable), class II (barely acceptable), and class III (not acceptable). Suppose that, of 2000 randomly selected cucumbers, there are 1200 in class I, 500 in class II, and 300 in class III. We may state that the probability that a cucumber is in class I is 1200/2000, or 3/5, and write P(I) = 0.6. Similarly P(II) = 0.25 and P(III) = 0.15.

Example 4.3 Suppose that we wish to calculate the probability that jars of preserves (made on a production line) are improperly labeled. We decide to take every 100th jar and classify it as properly or improperly labeled. If we look at 1000 jars, so that n = 1000, and find 73 mislabeled, so that r = 73, then the probability that a jar will be mislabeled is 0.073. We are, of course, assuming that n is large enough for the ratio r/n to have stabilized.

4.3 Some Laws of Probability

After we have calculated some elementary probabilities, we need basic rules for calculating probabilities of more complicated events. To illustrate this, reconsider example 4.2, and suppose that we know that the probability of a cucumber being in class I is 0.6 and, further, that the probability that a cucumber is longer than 50 cm is 0.8. How would we calculate the probabilities that (i) a cucumber is in class I *and* is over 50 cm, (ii) a cucumber is in class I *or* is under 50 cm, (iii) a cucumber is neither in class I *nor* is over 50 cm? To answer questions of this kind we shall develop four basic laws of probability. Before starting, we need to know that when we say that two events A and B are *mutually exclusive,* we mean that A and B cannot occur simultaneously.

Law 1 If two events A and B are mutually exclusive, then the probability that A *or* B occurs is the probability of A plus the probability of B. We write P(A or B) = P(A) + P(B). One justifies this law by supposing that in n trials, event A occurs r_A times and event B occurs r_B times, so that A or B occurs $r_A + r_B$ times. Thus P(A or B) $= (r_A + r_B)/n = r_A/n + r_B/n$. But r_A/n is P(A) and r_B/n is P(B).

Law 2 If two events A and B are not mutually exclusive (i.e., they can occur simultaneously), then P(A or B) = P(A) + P(B) − P(A and B).

To justify Law 2, suppose that of the n trials, A occurs r_A times, B occurs r_B times and A and B occur r_{AB} times. The "or" used here is the inclusive "or," so that A occurs (by itself) $r_A - r_{AB}$ times, B occurs (by itself) $r_B - r_{AB}$ times, and A or B occur $(r_A - r_{AB}) + (r_B - r_{AB}) + r_{AB}$ times which, when divided by n, gives P(A) + P(B) − P(A and B).

Example 4.4 The probability that (A), a potato weighs more than 0.1 kg is 0.3 and (B), that the potato is acceptable to a buyer is 0.8. Further, if the probability that the potato weighs more than 0.1 kg *or* is acceptable to a buyer is 0.9, we can find the probability that a potato weighs more than 0.1 kg and is acceptable to a buyer from Law 2 as follows:

$$P(\text{A and B}) = P(A) + P(B) - P(\text{A or B}) = 0.3 + 0.8 - 0.9 = 0.2$$

Laws 1 and 2 can be illustrated by the Venn diagrams of figures 4.1.1 and 4.1.2, where the probability of an event A is the proportion of total area inside circle A. In figure 4.1.1, A and B are mutually exclusive; the two circles representing events A and B do not intersect, whereas in figure 4.1.2, the area of intersection represents the probability of A and B occurring.

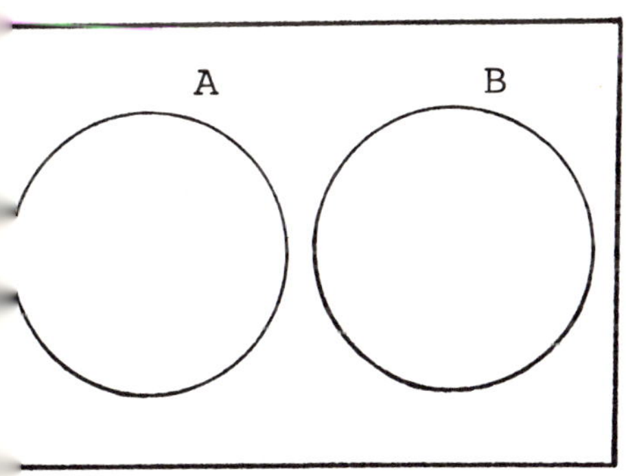

Fig. 4.1.1 *Two mutually exclusive events.*

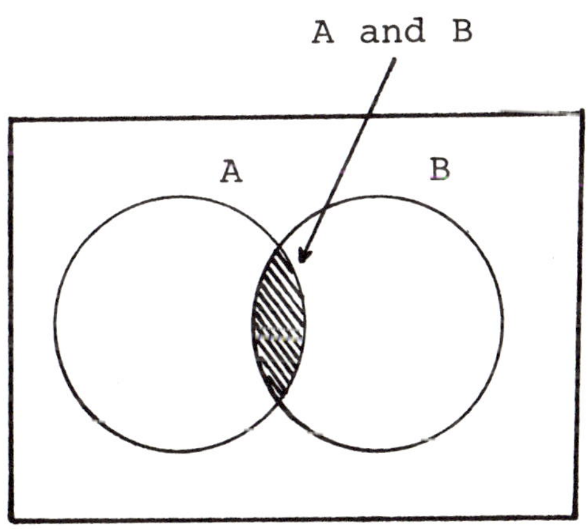

Fig. 4.1.2 *Two nonmutually exclusive events.*

For Laws 3 and 4 it is necessary to make a distinction between *dependent* and *independent* events. In an intuitive sense two events are independent if they do not affect each other. Thus, the yield per acre of broccoli in Ontario and the average weight of Florida oranges can be regarded as independent events, whereas the average weight of Florida oranges and the number of days of frost per year in Florida are dependent events. For two events, A and B, the probability that A will occur given that B has occurred is called the *conditional probability* of A given B and is written P(A|B).

Law 3 If two events A and B are dependent, then the conditional probability of A given B is the probability that A and B occur divided by the probability of B. We write P(A|B) = P(A and B)/P(B). Similarly P(B|A) = P(B and A)/P(A).

Law 3 can be justified by looking at figure 4.1.2. If B is known to have occurred, we are only concerned with the area inside the B circle. The probability of A given B is the proportion of the area inside B that is also inside A, that is, P(A and B) divided by P(B).

Law 4 If two events A and B are independent, the probability of the occurrence of A *and* B simultaneously is the probability of A times the probability of B. We write P(A and B) = P(A)P(B).

The justification for Law 4 follows from Law 3, where, if A and B are independent, then intuitively, P(A|B) = P(A), so that P(A) = P(A and B)/P(B) or P(A and B) = P(A)P(B).

Example 4.5 In example 4.4, use Law 4 to see if the two events "a potato weighs more than 0.1 kg" and "a potato is acceptable to a buyer" are dependent or independent. If they were independent, then P(A and B) = P(A)P(B). But P(A and B) = 0.2, P(A) = 0.3, and P(B) = 0.8. Thus P(A and B) ≠ P(A)P(B) and the two events are not independent but dependent.

Example 4.6 From example 4.4, find the probability that a potato is acceptable to a buyer knowing that it weighs more than 0.1 kg, that is, find P(B|A). Now from Law 3, P(B|A) = P(A and B) ÷ P(A) or 0.2/0.3 = 2/3.

In the next example we illustrate the use of the laws of probability in a common type of problem where two events occur sequentially in time.

Example 4.7 A box contains 12 jars of salad dressing of which 4 were produced on Monday and 8 on Tuesday. If 2 jars are selected consecutively, what is the probability that 1 was made on Monday and 1 on Tuesday? Using the letters M and T for Monday and

Tuesday, we need to find P(M and T). This is the probability that the first is M and the second is T *or* the first is T and the second is M. Since these two statements (events) are mutually exclusive, by Law 1 we write

P(M and T) = P(first M and second T) + P(first T and second M)

By Law 3

P(first M and second T) = P(firstM)P(second T | first M)

P(first T and second M) = P(firstT)P(second M | first T)

But, P(first M) = 4/12, P(second T|first M) = 8/11

Thus, P(M and T) $= \frac{4}{12} \cdot \frac{8}{11} + \frac{8}{12} \cdot \frac{4}{11} = .4848.$

4.4 Permutations and Combinations

In example 4.1 we can easily calculate the probability of picking a head lettuce when one lettuce is picked from the consignment. But suppose that we pick two or more lettuces. How do we calculate the probability of 5 head lettuces when 10 lettuces are picked? To answer questions of this type we proceed as follows. For a set of n different or distinguishable objects, an arrangement of r of them is called a *permutation* of size r. The total number of permutations of size r from n is denoted by p_r^n and calculated as

$$p_r^n = \frac{n!}{(n - r)!}$$

where n! (read n factorial) is n(n − 1)(n − 2) . . . 1 and (n − r)! = (n − r)(n − r − 1) . . . 1. By convention 0! = 1.

Example 4.8 How many different ways are there of arranging 5 brands of ketchup on a shelf if the shelf can hold (i) exactly 2 bottles at a time, (ii) all 5 bottles together? Calling the 5 brands A, B, C, D, E, some of the arrangements of size 2 are AB, AC, AD, AE, BA, BC, ... One should be able to verify that there are 20 permutations, agreeing with the formula

$$p_2^5 = \frac{5!}{(5 - 2)!} = \frac{120}{6} = 20$$

In part (b), n = 5 and r = 5 so that p_5^5 = 5!/0! = 120. Note in part (a) that the arrangement AB is considered different from BA.

Example 4.9 In example 4.8(i), what is the probability that brands A and E will be placed side by side? Of the 20 possible permutations, 2 have A and E side by side (AE and EA). Thus, the probability is 2/20 or 0.1. For a set of n objects, a subset of r of them (chosen without regard to their order of selection) is called a *combination* of size r. The number of different combinations of size r from n is denoted by $\binom{n}{r}$ and calculated as

$$\binom{n}{r} = \frac{n!}{r!(n - r)!}$$

Example 4.10 Continuing example 4.8, suppose that we do not wish to distinguish between AB and BA. Then we can ask the number of combinations of size 2. The answer is

$$\binom{5}{2} = \frac{5!}{2!3!} = \frac{120}{(2)(6)} = 10$$

We can verify the answer by writing out all 10 combinations.

If we have n repeated trials, each with 2 possible outcomes (e.g., defective and nondefective) the possible number of different arrangements each having x defectives and n − x nondefectives is $\binom{n}{x}$.

Example 4.11 Five people are asked to test a drink and answer yes or no to the question "Is the taste acceptable?". How many ways are there to get 2 "nos" and 3 "yesses"?

This is a case of n = 5 trials with x = 2, n − x = 3. The number of different arrangements is $\binom{5}{2}$ = 10. To justify the answer we can write out all ten arrangements.

Person				
1	2	3	4	5
N	N	Y	Y	Y
N	Y	N	Y	Y
N	Y	Y	N	Y
N	Y	Y	Y	N
Y	N	N	Y	Y
Y	N	Y	N	Y
Y	N	Y	Y	N
Y	Y	N	N	Y
Y	Y	N	Y	N
Y	Y	Y	N	N

4.5 Random Variables and Distributions

Suppose that, after examining a large population of lemons for scale spots, we obtain the following information. P(no scale spots) = 0.4, P(one scale spot) = 0.3, P(two scale spots) = 0.2, P(three scale spots) = 0.1. It would be convenient to let x represent the number of scale spots; then, writing P(x = 0) in place of P(no scale spots), or, even more simply, writing p(0), we have

$$\begin{aligned} p(0) &= 0.4 \\ p(1) &= 0.3 \\ p(2) &= 0.2 \\ p(3) &= 0.1 \end{aligned} \qquad \text{and} \qquad \sum_{\text{all } x} p(x) = 1$$

Note that the sum of the probabilities for all possible values of x add to 1. Since x is a variable (its value varies from lemon to lemon) whose values have attached probabilities, it is called a *random variable.* The values that x can assume and their respective probabilities are called the *distribution* of x. In example 4.1, if x is the number of head lettuces obtained when we pick 1 lettuce from the consignment, then x can take the values 0 or 1 with probabilities 1/3 and 2/3, respectively. The distribution of x is

$$\begin{aligned} p(0) &= 1/3 \\ p(1) &= 2/3 \end{aligned}$$

Returning to the discussion of the number of scale spots on lemons, suppose that there are N lemons in the population of which 0.4N have no spots, 0.3N have one spot, etc. The total number of spots is 0.4N(0) + 0.3N(1) + 0.2N(2) + 0.1N(3) which, when divided by N, gives the average number of spots, μ, per lemon in the population. Thus,

$$\mu = (0.4)(0) + (0.3)(1) + (0.2)(2) + (0.1)(3) = 1.0$$

Note that $\mu = \sum_{x=0}^{3} x\, p(x)$. This formula is quite general. To find the population average value of x, sometimes referred to as the *expected value* of x and denoted by E(x) or μ, multiply each x by its probability and sum, i.e.,

$$\mu = E(x) = \sum_{\text{all } x} x\, p(x)$$

To measure the variability in the values of x, we calculate the *population variance* of x, denoted by σ^2 or var(x) as follows:

$$\text{Var}(x) = \sigma^2 = \sum_{\text{all } x} (x - \mu)^2\, p(x)$$

Note that variance is measured from the expected value. The differences are squared to eliminate negative signs, and then multiplied (or weighted) by the probabilities of x. For the lemon data

$$\sigma^2 = (0 - 1)^2\,(0.4) + (1 - 1)^2\,(0.3) + (2 - 1)^2\,(0.2) + (3 - 1)^2\,(0.1) = 1.0$$

It is imperative that the student not confuse the sample mean and variance $\overline{X}$ and S^2 with the expected value and the population variance denoted by μ and σ^2 calculated from the population probabilities.

Finally we state (but not prove) the expected value and variance of the difference Y of two random variables X_1 and X_2, where $E(X_1) = \mu_1$, $E(X_2) = \mu_2$ and σ_1^2 and σ_2^2 are their respective variances.

$$E(Y) = E(X_1 - X_2) = \mu_1 - \mu_2$$
$$\text{Var}(Y) = \text{Var}(X_1 - X_2) = \sigma_1^2 + \sigma_2^2$$

Exercises

1. If 100 out of 800 cucumbers are longer than 40 cm, what is the probability that (i) a single randomly selected cucumber will be longer than 40 cm, (ii) shorter than 40 cm? What is the probability that two randomly selected cucumbers will be longer than 40 cm if (i) the first is replaced before the second is selected, (ii) the first is not replaced?

2. If the letters of the word "statistics" are arranged at random, what is the probability that
   (a) A and C are next to each other
   (b) the arrangement will start with a T
   (c) the arrangement will end with a T
   (d) the arrangement will start and end with a T?

3. A processor has two machines, the first of which is idle 40% of the time and the second of which is idle 30% of the time. Assuming that both machines are idle independently of each other, what is the probability that
   (a) both will be idle
   (b) neither will be idle
   (c) at least one will be idle?

4. A farmer is bidding for two pieces of land, A and B. The probability of his getting piece A is 0.7 and of getting piece B is 0.2. What is the probability of his getting
   (a) both pieces
   (b) at least one piece
   (c) neither piece?

5. In exercise 4, if the probability of his getting piece A given that he gets B is 0.9, find the probability of his getting
   (a) pieces A and B
   (b) piece B given that he gets A

6. Let A and B be two events for which $P(A) = 1/2$, $P(B) = 1/3$, $P(A|B) = 1/7$. (a) Find P(A) and P(B). (b) Find $P(B|A)$. (c) Are A and B independent? Why?

7. In a standard deck of cards, a random sample of two cards is drawn. How many different pairs are possible? How many of the possible pairs could contain a king?

8. At a particular university, 40% of the students are studying agriculture, 55% of the students are male, and 30% of the students are male and studying agriculture. Find the probability that (i) a student is male given that he/she is studying agriculture, (ii) a student is studying agriculture given that she is female.

9. Rework example 4.6 using the idea of section 4.4.

10. Of 20 students, 12 are male and 8 are female. A committee of 4 is randomly selected. What is the probability that it will contain (i) two females, (ii) at least two females?

11. Suppose that a large truckload of California strawberries arrives in New York and that 20% of the cases are already overripe. A random sample of 15 cases is examined. What

is the probability of (i) 3 overripe cases being found, (ii) 3 or fewer overripe cases being found?

12. In exercise 3, let X be the number of idle machines (so that X = 0 or 1 or 2). Find the distribution of X. Then find the expected value and variance of X.

13. In exercise 1 where two cucumbers are selected, if X represents the number of cucumbers longer than 40 cm, find the distribution of X, its mean, μ, and its variance if (i) the first is replaced before the second is picked, (ii) the first is not replaced.

14. In exercise 4, if X is the number of pieces of land the farmer gets, find the distribution of X. Then find its expected value and variance.

15. In exercise 7, find the distribution for the number of kings in a drawing of a pair of cards, without replacement. Then find the expected value and variance of X.

16. In a fruit-canning process, if the weight of the cans is a random variable, X_1, for which $\mu_1 = 60$ gm and $\sigma_1^2 = 9$ gm^2 and the can and its contents are another random variable, X_2, for which $\mu_2 = 500$ gm and $\sigma_2^2 = 100$ gm^2, find the expected value and variance of the contents.

5.

Four Important Distributions

In chapter 4 it was mentioned that, in an experiment, we are usually concerned with a particular random variable x (such as the number of scale spots on a lemon or the weight of a load of potatoes) and that the distribution of the random variable is known or assumed to be known. From this one can proceed to some of the central problems in statistics, such as estimating population parameters from sample statistics using probability. In this chapter we concern ourselves with four particular distributions which occur frequently enough to make their study worthwhile. Three of these distributions, the binomial, the Poisson, and the hypergeometric distributions, deal with discrete data (see section 1.2) where the random variables are counted and the fourth, the normal distribution, deals with continuous or measured data.

5.1 The Hypergeometric Distribution

Suppose that a population with which we are concerned has N objects of which X are defective and N − X are acceptable. From this population we take a random sample (section 1.1) of n objects. We would like to calculate the probability that there are x defectives in the sample, denoted by p(x).

We know from section 4.4 that the total possible number of different samples of size n from population N is $\binom{N}{n}$ (see example 4.10). Further, the total possible number of ways (combinations) of choosing x defectives from X is $\binom{X}{x}$ and the total possible number of ways of choosing n − x nondefectives from N − X is $\binom{N-X}{n-x}$. Thus the number of ways of choosing x from X and n − x from N − X is $\binom{X}{x}\binom{N-X}{n-x}$. Using the classical definition of probability, the probability of x defectives and n − x nondefectives is

$$p(x) = \frac{\binom{X}{x}\binom{N-X}{n-x}}{\binom{N}{n}} \qquad \text{for } x = 0, 1, 2, \ldots, n$$

This is known as the *hypergeometric distribution.*

We can show (Malik and Mullen 1971) that the expected value of the random variable is $\frac{nx}{N}$ and that its variance is

$$\frac{N-n}{N-1} \cdot n \cdot \frac{X}{N}\left(1 - \frac{X}{N}\right).$$

Example 5.1 A box contains 20 apples of which 4 are spoiled. A random sample of 5 is chosen. What is the probability that it contains (i) 2 spoiled ones, (ii) more than 2 spoiled ones?

With N = 20, X = 4, n = 5, x = 2, we have

$$p(2) = \frac{\binom{4}{2}\binom{16}{3}}{\binom{20}{5}} = 0.2167$$

P(more than 2) = p(2) + p(3) + p(4) = 0.2487
If all possible samples of size 5 were taken, the expected value of x would be 1 and the variance of x would be 12/19.

5.2 The Binomial Distribution

For the hypergeometric distribution, if N is very large we may suppose that as each successive member of the sample is chosen, the probability that the next member is defective does not change. For example, if a population contains 20 bottles of sauce of which 2 are defective, the probabilities of 2 defectives in a row are 2/20 (.1) and

1/19 (:05), which are very different, whereas if there are 10000 bottles in the population of which 1000 are defective, the probabilities of 2 successive defectives are 1000/10000 (.1) and 999/9999 (.0999), which are about equal.

To calculate probabilities p(x) for the number of defectives, x, in a sample of size n from a very large (or essentially infinite) population, we use the *binomial distribution* as follows:

$$p(x) = \binom{n}{x} p^x (1 - p)^{n-x} \qquad \text{for } x = 0, 1, 2, \ldots, n$$

where p is the probability of a defective being chosen at each trial. (Note that p remains constant throughout the sampling.)

We may justify the binomial distribution as follows. Each member of the sample of size n is either defective or nondefective and the number of different samples with x defectives is $\binom{n}{x}$. Any particular one of the $\binom{n}{x}$ samples has x defectives, each chosen with probability p, and n − x nondefectives, each with probability 1 − p, for a total probability of $p^x (1 - p)^{n-x}$. For all possible samples the probability is $\binom{n}{x} p^x (1 - p)^{n-x}$. This is true for x = 0 or 1 up to x = n. The expected value of a binomially distributed random variable can be shown to be np and its variance np(1 − p).

Example 5.2 Sauce is being bottled and labeled on a production line. To test the labeling, 4 bottles are selected randomly every hour. It is known from past experience that the process is 20% defective (i.e., 1 bottle in 5 is mislabeled). What is the probability that 3 of the 4 bottles will be defective?

It is appropriate to use the binomial distribution, since the population (number of bottles labeled per hour) is presumed large. Here, n = 4, x = 3, p = 0.2, so that

$$p(3) = \binom{4}{3} (0.2)^3 (0.8)^1 = 0.0256$$

This would generally be regarded as a small probability. Thus, if we did find 3 defectives in a sample of size 4, we would conclude that, since the probability of its happening by chance was so small (almost impossible), that p, the chance of a bottle being mislabeled, was really greater than 0.2, or that the percentage of mislabelings had increased.

To avoid the tedium of calculations, values of $\binom{n}{x}$ are given in

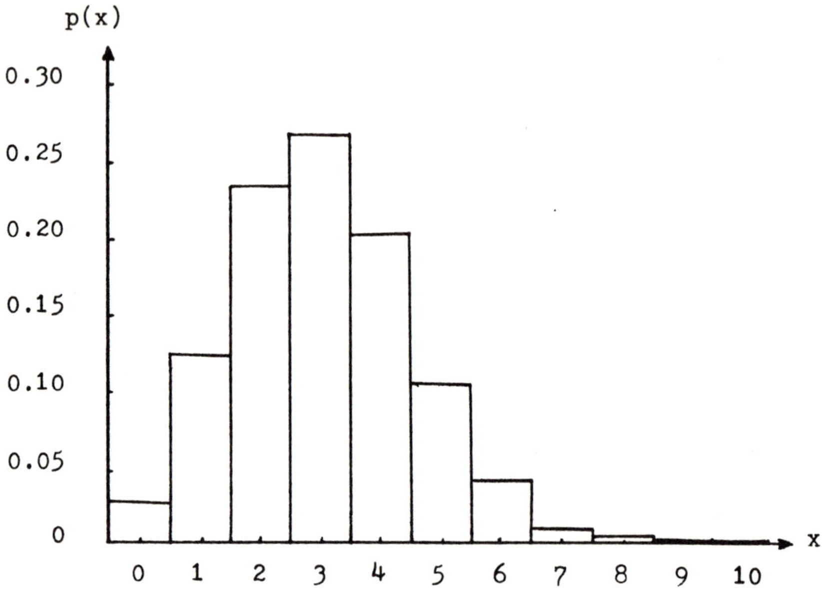

Fig. 5.1 *The binomial distribution for n = 10, p = 0.3.*

Appendix Table 1 for values of n from 0 to 20. For example 5.2 we may read directly from Table 1 that $\binom{4}{3} = 4$, so that p(3) = 0.0256.

It is sometimes convenient or desirable to represent binomial probabilities graphically. This may be simply done as illustrated in figure 5.1 for n = 10, p = 0.3.

5.3 The Poisson Distribution

If we are dealing with a binomial random variable for which n is large and p is small, the probabilities of x, p(x), can be more conveniently calculated using the *Poisson distribution* for which

$$p(x) = \frac{e^{-\lambda}\lambda^{x}}{x!} \qquad \text{for } x = 0, 1, 2, \ldots$$

where $\lambda = np$. Values of $e^{-\lambda}$ can be found in Appendix Table 2. The approximation of binomial probabilities using the Poisson distribution is generally adequate if n is larger than 20 and p is less than 0.05. If n is larger than 100, then p may be as large as 0.1. Because of the small values of p, the Poisson distribution is sometimes known as the distribution of rare events. The expected value of a Poisson-distributed random variable is λ and its variance is also λ.

Example 5.3 Reconsider example 5.2 and suppose that the proportion of defectives is known (or conjectured) to be 0.05. Suppose that every hour a random sample of 40 bottles is taken. What is the probability of drawing (i) 1 defective bottle, (ii) 3 or more defective bottles?

This is a binomial problem, but n is large and p is small so that we may use the Poisson distribution with $\lambda = np = 2$.

Using Appendix Table 2, $e^{-2} = 0.135$, so that

$$p(1) = \frac{e^{-2}\,2^1}{1!} = 0.2707$$

The probability of 3 or more defectives equals 1 minus the probability of 0, 1, or 2. Thus

$$P(3 \text{ or more defectives}) = 1 - \frac{e^{-2}\,2^0}{0!} - \frac{e^{-2}\,2^1}{1!} - \frac{e^{-2}\,2^2}{2!} = 0.3233$$

5.4 The Normal Distribution

The hypergeometric, binomial, and Poisson distributions all deal with discrete or counted data. In each case the random variable x can take on only a limited number of values. For instance, in the binomial distribution x can be 0, 1, 2, . . . , n. It cannot be negative, nor can it assume noninteger values. The most important distribution for dealing with continuous or measured data is called the *normal distribution,* whose formula is

$$p(x) = \frac{1}{\sigma\sqrt{2\pi}}\, e^{-\frac{1}{2}\left(\frac{x-\mu}{\sigma}\right)^2} \quad \text{for all values of x,}$$

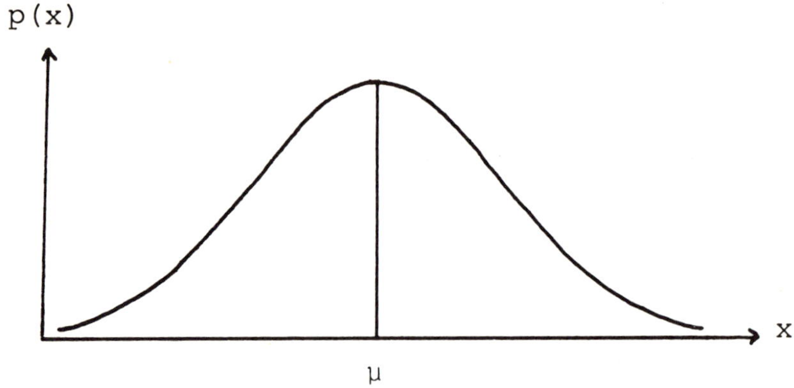

Fig. 5.2 *Typical normal distribution.*

where μ is the expected value of the distribution, and σ is the population standard deviation. The graph of a typical normal distribution is given in figure 5.2.

The value of σ controls the width or spread of the distribution. To use the normal distribution one does not calculate p(x) for a particular value of x. Rather, one calculates the probability that x lies in an interval by finding the area under the normal distribution within the required interval. Thus, probability for the normal distribution is synonymous with area. This means that the total area under the normal distribution (or normal curve) is unity. Although in mathematics, area is associated with the integral calculus, we shall show how to find areas (and thus probabilities) using Appendix Table 3, the table of normal probabilities. Since the expected value of a normal random variable is μ and its variance is σ^2, one may show that $(x - \mu)/\sigma$ has expected value zero and variance (standard deviation) one. This is of particular importance, as we shall see. There are, in theory, an infinite number of different normal distributions. For example, the weights of 10-week-old chickens might have one normal distribution, with a particular mean μ and variance σ^2, while the heights of corn stalks might have another normal distribution, and circumference of apples from Ontario, a third. Clearly, we cannot have a table of normal distribution probabilities for each distribution, and, fortunately, it is not necessary to do so. For any normally distributed random variable, we may subtract its expected value μ and divide by its standard deviation. The resulting new random variable, called z, has expected value zero and variance one and has the *standard normal distribution*. Thus,

$$z = \frac{x - \mu}{\sigma}$$

has a standard normal distribution with $\mu = 0$ and $\sigma = 1$. The curve of the standard normal distribution is centered at zero and has the bulk of its area between -3 and $+3$ as shown in figure 5.3.

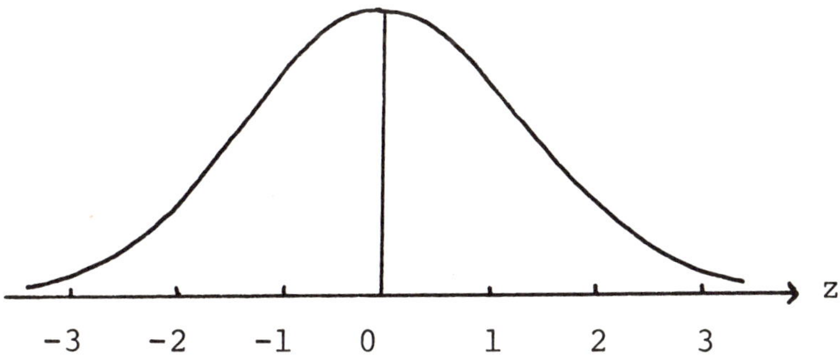

Fig. 5.3 *Standard normal curve.*

The probability that z, which has a standard normal distribution, lies between any two numbers a and b (we assume that $a \leq b$) is read from Appendix Table 3.

Example 5.4 Using the standard normal distribution of Appendix Table 3, find the probability that z is (i) between 0 and 1, (ii) between -1 and 0, (iii) between -1 and 1, (iv) between -2 and -1, (v) greater than 1, (vi) equal to 1, (vii) greater than 4, (viii) less than -4.

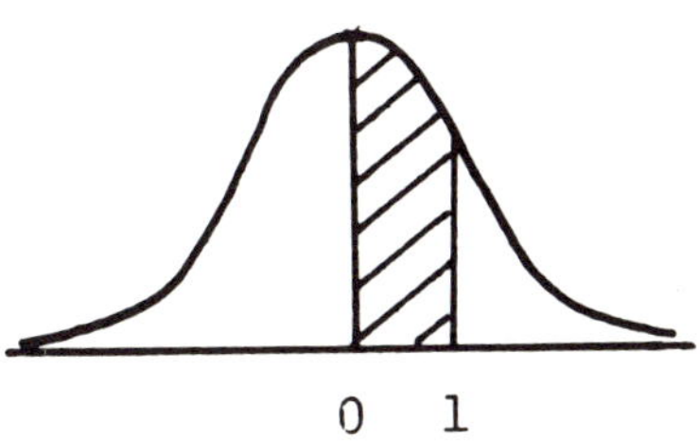

Fig. 5.4.1

The probability of z being between 0 and 1 is found by reading the A (for area) value for $z = 1$. This corresponds to the shaded area and is $A = 0.3413$. Thus the probability that z lies between 0 and 1 is 0.3413 or $P(0 < z < 1) = 0.3413$.

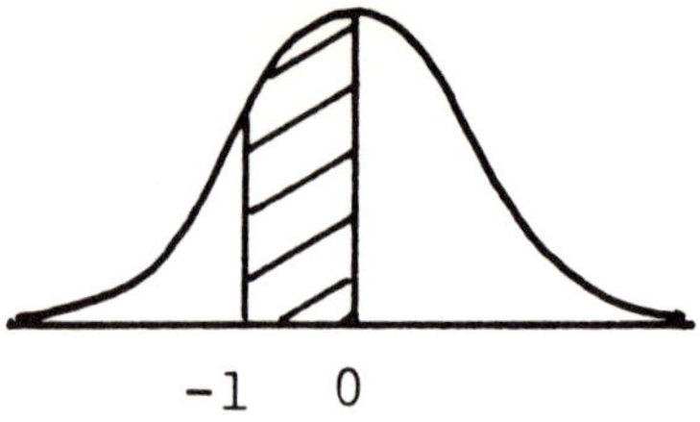

Fig. 5.4.2

From the symmetry of the normal distribution, the area between −1 and 0 is the same as between 0 and 1. Hence $P(-1 < z < 0) = 0.3413$.

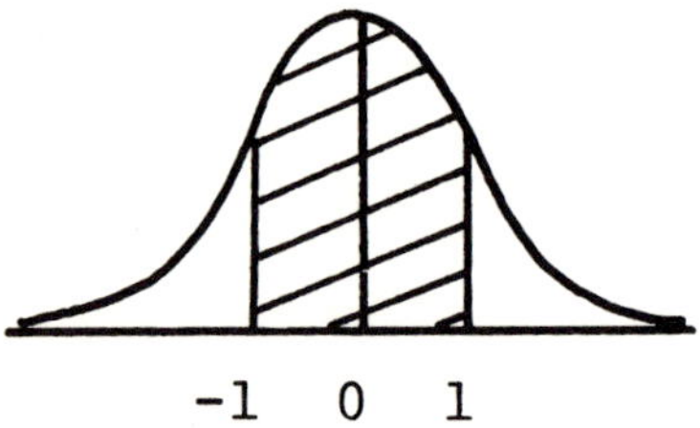

Fig. 5.4.3

The area between −1 and 1 is the sum of the areas between −1 and 0 and 0 and 1. Thus $P(-1 < z < 1) = 0.6826$.

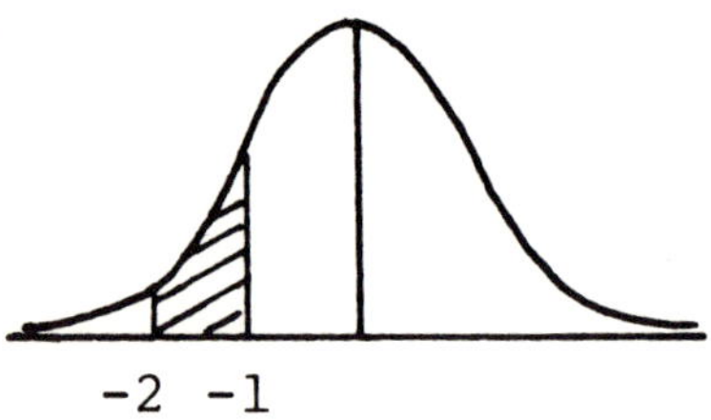

Fig. 5.4.4

The area between −2 and −1 is, by symmetry, the same as between 1 and 2. But the area between 1 and 2 is the area between 0 and 2 minus the area between 0 and 1. Thus, read A for $z = 2$ and subtract A for $z = 1$. $P(-2 < z < -1) = 0.4773 - 0.3413 = 0.1360$.

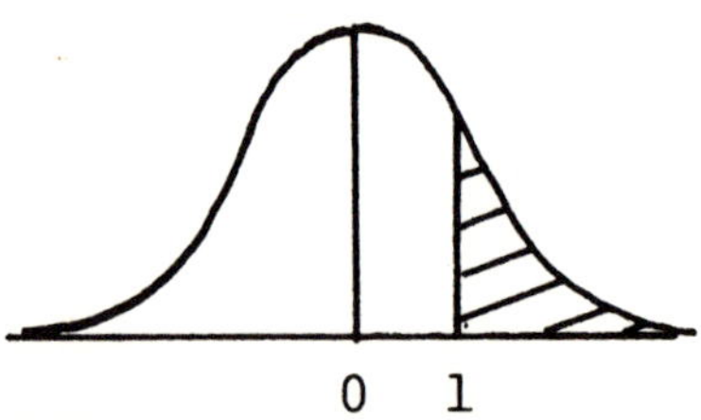

Fig. 5.4.5

The area greater than 1 is 0.5 minus the area between 0 and 1, so that $p(z > 1) = 0.5 - 0.3413 = 0.1587$.

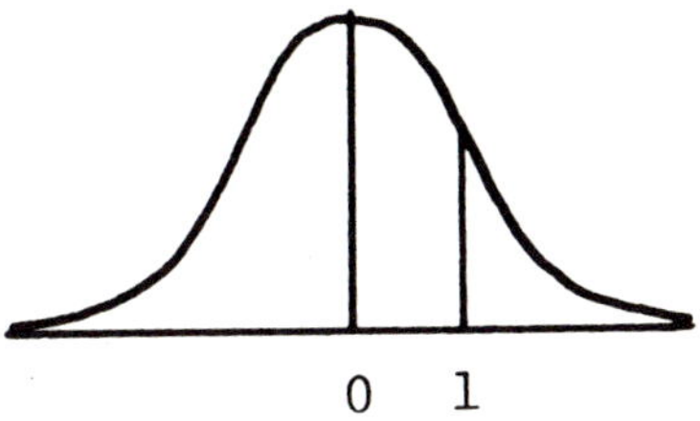

Fig. 5.4.6

The area at z = 1 is area between 1 and 1, i.e., it is zero. Thus, P(z = 1) = 0.

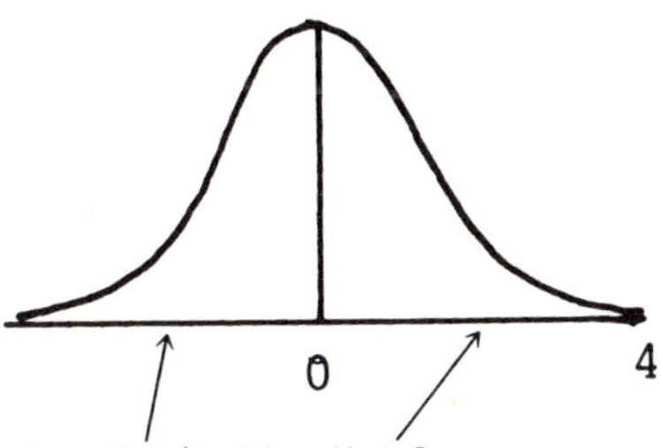

Fig. 5.4.7, Fig. 5.4.8

The area greater than z = 4 is negligible because the area between 0 and 3 is 0.4987. Thus, P(z > 4) = 0.0.
Similarly, P(z < −4) = 0.0.

We illustrate, with the following example, how one may use the standard normal distribution to calculate probabilities for any normal distribution.

Example 5.5 Assuming that fat content in milk has a normal distribution with mean $\mu = 3.5$ (%) and standard deviation $\sigma = 0.3$ (%), what is the probability that the fat content of a randomly chosen sample of milk will be between 2.9 and 3.8?

$$\text{When } x = 2.9,\ z = \frac{2.9 - \mu}{\sigma} = \frac{2.9 - 3.5}{0.3} = -2.0$$

$$\text{When } x = 3.8,\ z = \frac{3.8 - \mu}{\sigma} = \frac{3.8 - 3.5}{0.3} = 1.0$$

The probability that milk fat content (x) is between 2.9 and 3.8 is the same as $P(-2 < z < 1)$. From Appendix Table 3, this is 0.8186.

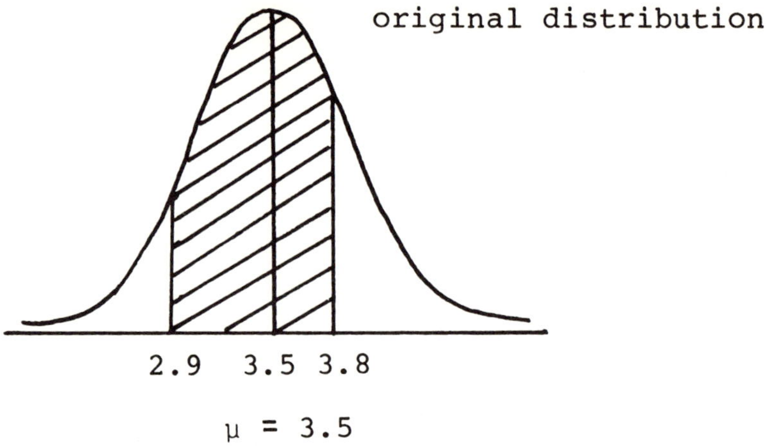

Fig. 5.5.1

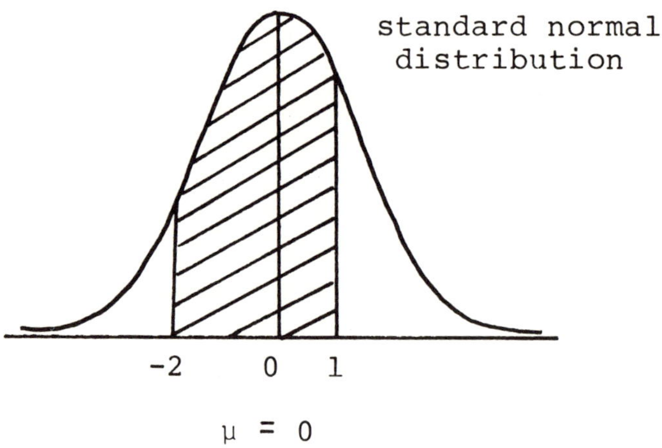

Fig. 5.5.2

Often, instead of writing "x has a normal distribution with mean μ and variance σ^2," one simply writes $x \sim N(\mu, \sigma^2)$. Thus, in example 5.5 we could write $x \sim N(3.5, 0.09)$.

A notation that will later prove convenient is the following: Z_α is that point on the standard normal distribution which has area α (or 100 α%) greater than it, i.e., $P[z > Z_\alpha] = \alpha$, as shown in figure 5.6.

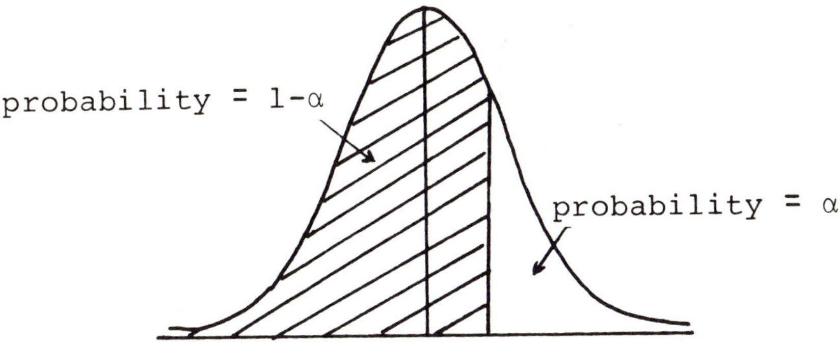

Fig. 5.6 *Showing Z_α.*

From the symmetry of the normal distribution $Z_{1-\alpha} = -Z_\alpha$. We can verify from Appendix Table 3 that $Z_{.05} = 1.65$, $Z_{.95} = -Z_{.05} = -1.65$, $Z_{.5} = 0$, and $Z_{.0} = 1.28$. The two points that contain the central $100(1 - \alpha)\%$ of the area are $\pm Z_{\alpha/2}$ as shown in figure 5.7.

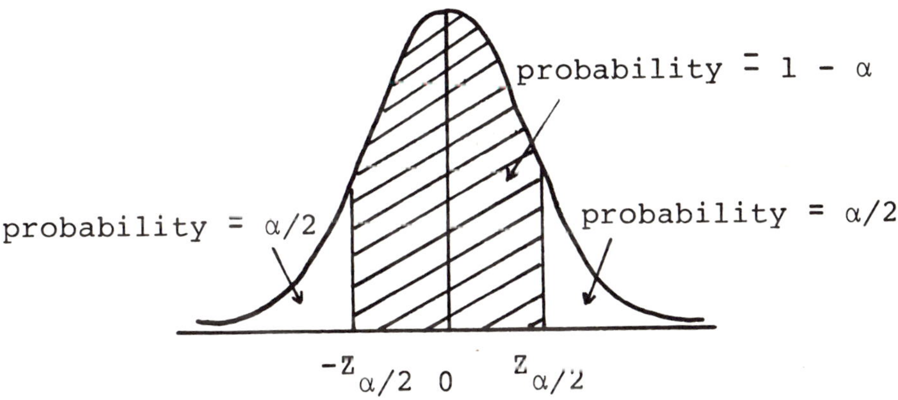

Fig. 5.7 *Central 100(1 − α)% area.*

5.5 The Normal Approximation to the Binomial Distribution

We saw in section 5.3 when calculating binomial probabilities that if n is large and p is small, one can use the Poisson distribution for convenience. Similarly, if n is large and p is moderate, one can use

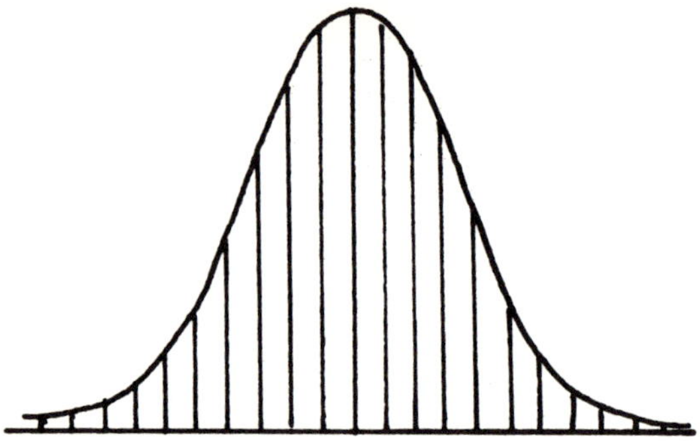

Fig. 5.8 *Binomial distribution (with $p = .5$, $n = 20$) encased by a normal distribution with mean 10 and variance 5.*

the normal distribution to obtain good approximations to binomial probabilities. The approximations are generally adequate because the binomial distribution is close to being symmetrical for moderate values of p (see figure 5.1), so that it may be "encased" in the symmetric normal distribution as shown in figure 5.8 (for which $n = 20$, $p = 0.5$).

It is clear from figure 5.8 that the approximation will be best if the normal distribution has the same mean and variance as the binomial distribution. From section 5.2 we know that the mean is np and the variance is $np(1 - p)$. Thus, to find the probability that the binomial random variable lies between 7 and 11 (inclusive) we could find $P(7 \le x \le 11)$ where x has a normal distribution with mean np and variance $np(1 - p)$.

Since the binomial distribution is discrete, and the normal distribution continuous, the approximation can be improved by using an *adjustment for continuity* of 0.5 subtracted from the lower value and added to the upper value. This is illustrated in example 5.6.

Example 5.6 If x has a binomial distribution with $n = 20$, $p = .4$, use the normal approximation to find (i) $P(5 \le x \le 10)$, (ii) $P(5 < x < 10)$, (iii) $P(x > 10)$.

The approximating normal distribution will have $\mu = np = 8$ and $\sigma^2 = np(1 - p) = 4.8$. We denote the normal random variable by y. Thus,

(a) $P(5 \le x \le 10) \cong$
$P(4.5 < y < 10.5) = P(-1.60 < z < 1.14) = 0.8181$

(b) $P(5 < x < 10) \cong P(5.5 < y < 9.5) = P(-1.14 < z < 0.68) = 0.6247$

(c) $P(x > 10) \cong P(y > 10.5) = P(z > 1.14) = 0.1271$

Exercises

1. A box contains 3 defective and 9 nondefective items. Three items are drawn out (without replacement). Let x be the number of defectives drawn. Write down p(x), the distribution of x. Calculate p(0) and p(1). What is the expected value and variance of x?

2. From a group of 5 women and 7 men, a panel of 4 people is selected. Let x be the number of women on the panel. What is the distribution of x? Calculate the probability that there will be (i) no women, (ii) 4 women on the panel.

3. In exercise 1, calculate p(0), p(1), p(2), and p(3). Then calculate μ and σ^2 by the methods of section 4.5. Verify the results using the formulas for μ and σ^2 of section 5.1.

4. In exercise 2, calculate p(0), p(1), p(2), p(3), and p(4). Then calculate μ and σ^2 by the methods of section 4.5. Verify the results using the formulas for μ and σ^2 of section 5.1.

5. In exercise 1, if each item is replaced before the next one is drawn, p, the probability of a defective being drawn, does not change during the experiment. Write down p(x), the distribution of x, and from it calculate p(0) and p(1).

6. If it is known that 15% of bottles labeled on a production line are mislabeled, find the probability that in a random sample of 20 bottles, 3, or fewer will be mislabeled. If this sampling procedure is repeated many times, what is the expected number of mislabeled bottles found?

7. A true-false examination with 15 questions, each having 1 true and 4 false answers, is given. If a student picks the answer entirely at random, what is the probability of getting (i) 4 answers correct, (ii) 2 answers correct, (iii) more than 2 answers correct?

8. If x has a binomial distribution with $n = 20$, $p = 0.05$, compare numerically the values of p(x) for $x = 0, 1, 2, \ldots, 10$ using (i) the binomial distribution, and (ii) the Poisson distribution.

9. If the probability of a rare trait (albinism, for example) is 0.001, and a random sample of 5000 individuals is chosen, find the probability of (i) 2, (ii) 5 individuals in the sample exhibiting that trait.

10. In exercise 1, find the probability of 2 defectives given that there is at least 1 defective.

11. In exercise 7, find the probability of getting five questions answered correctly, given that the student answers at least 3 correctly.

12. In exercise 9, find the probability of choosing 5 individuals given that at least 2 have the trait.

13. Find the area under the standard normal distribution
    (a) between 0 and 0.8
    (b) between -0.8 and 0.5
    (c) between -1.3 and 2.4
    (d) between 0.5 and 2.4
    (e) less than -1.3
    (f) greater than 3.8

14. Using the standard normal distribution, find the value of Z for which the
    (a) area between 0 and Z is 0.3849
    (b) area to the right of Z is 0.2742
    (c) area to the left of Z is 0.0808
    (d) area between -1.3 and Z is 0.7375

15. Using the standard normal distribution, find
    (a) $P(z > 1.28)$
    (b) $P(-2 < x < 1.5)$
    (c) $P(z > 4.0)$
    (d) $P(-2.3 < z < -1.4)$
    (e) $P(z < 1.7)$
    (f) $P(1.96 < z < 1.96)$

16. If the egg weight x of white leghorns is assumed to have a normal distribution with mean μ equal to 50 gm and standard deviation σ equal to 5 gm, find
    (a) $P(x > 50)$
    (b) $P(x > 52)$
    (c) $P(48 < x < 55)$
    (d) $P(x < 47)$
    (e) $P(x > 52 | x > 51)$

17. If it is known that 40% of a particular kind of cucumber have blemishes and a sample of 30 is selected, use the normal approximation to the binomial to find the probability that
    (a) 10 or fewer have blemishes
    (b) exactly 15 have blemishes
    (c) between 10 and 18 (inclusive) have blemishes

6.

Statistical Estimation

Chapters 1, 2, and 3 dealt with organizing, summarizing, and presenting sample data. Chapter 4 introduced probability, and chapter 5 dealt with four important probability distributions. We are now ready to use sample data and probability to discuss one of the fundamental problems of statistics, namely, how to infer from sample statistics to population parameters. Recall that parameters describe the population and are generally of great interest to experimenters. They obtain data in the form of samples, calculate the pertinent statistics, and then use these statistics to gain knowledge about the parameters. We shall discuss how this is done after looking more closely at two particular statistics of paramount importance, namely, the sample mean $\overline{X}$ and the sample variance S^2.

6.1 The Distribution of the Sample Mean and the Sample Variance

If a population contains N elements and a random sample of size n is chosen from it, we know from chapter 4 that there are $\binom{N}{n}$ possible samples of which we have but one. Each of the possible samples has a mean, median, variance, etc. In particular, there are $\binom{N}{n}$ possible sample means $\overline{X}_1$, $\overline{X}_2$, $\overline{X}_3$, etc. The totality of these sample means is called the *distribution of the sample mean.* This distribution of the sample mean, like any other distribution, has a mean or expected value, a variance, and a standard deviation. Suppose that the original population from which the sample was taken has a mean μ and a variance σ^2. Let us call the mean and variance of the distribution of the sample mean $\mu_{\overline{x}}$ and $\sigma^2_{\overline{x}}$, respectively. Then it can be shown that

$$\mu_{\overline{x}} = \mu \text{ and } \sigma^2_{\overline{x}} = \frac{\sigma^2(N - n)}{n(N - 1)}$$

For large populations, the formula for the variance is approximately

$$\sigma^2_{\overline{x}} = \frac{\sigma^2}{n}$$

We illustrate these results for a small population (N = 3) in example 6.1, but first state two important points. If the original population is normal, then the distribution of the sample mean is also normal. If the original distribution is not normal, a very important theorem in mathematical statistics called the *central limit theorem* tells us that the distribution of $\overline{X}$ is approximately normal, and that the approximation improves as the sample size n becomes larger.

Example 6.1 Consider a population of milk fat percentages with 4 observed values (N = 4): 3.0, 3.2, 2.8, 3.4 for which (the student should calculate this) $\mu = 3.1$, $\sigma^2 = 0.05$. The six possible samples of size 2 with their means are shown as follows:

Sample	Observed values	Sample mean ($\overline{X}$)
1	3.0, 3.2	3.1
2	3.0, 2.8	2.9
3	3.0, 3.4	3.2
4	3.2, 2.8	3.0
5	3.2, 3.4	3.3
6	2.8, 3.4	3.1

The six sample means form the distribution of sample means.

Their mean value is

$$\mu_{\bar{x}} = \frac{3.1 + 2.9 + \ldots + 3.2}{6} = 3.1 \qquad (\text{or } \mu_{\bar{x}} = \mu)$$

and their variance is

$$\sigma^2_{\bar{x}} = \frac{(3.1 - 3.1)^2 + (2.9 - 3.1)^2 + \ldots + (3.1 - 3.1)^2}{6} = .01667$$

$$\left(\text{or } \sigma^2_{\bar{x}} = \frac{\sigma^2}{n}\frac{(N - n)}{(N - 1)}\right)$$

Because the distribution of $\overline{X}$ is approximately normal, we may calculate probabilities for $\overline{X}$ using the ideas of normal distribution from section 5.4.

Example 6.2 Suppose that the population of milk fat content percentages is assumed to be approximately normal, with mean μ equal to 3.5 and standard deviation σ equal to 0.3. The distribution of sample means of samples of size 25 will have mean $\mu_{\bar{x}}$ equal to 3.5 and standard deviations $\sigma_{\bar{x}}$ equal to 0.3/5 or 0.06. To find the probability that $\overline{X}$ lies between 3.41 and 3.56, we calculate the corresponding standard normal deviates as follows:

$$\text{when } \overline{X} = 3.41,\ z = \frac{3.41 - \mu_{\bar{x}}}{\sigma_{\bar{x}}} = \frac{3.41 - 3.5}{0.06} = -1.5$$

$$\text{when } \overline{X} = 3.56,\ z = \frac{3.56 - \mu_{\bar{x}}}{\sigma_{\bar{x}}} = \frac{3.56 - 3.5}{0.06} = 1.0$$

The probability that mean milk fat content ($\bar{x}$) is between 3.41 and 3.56 is the same as $P(-1.5 < z < 1.0)$. From Appendix Table 3, this is 0.7745.

In subsequent work, it will be necessary to consider the difference of two means, $\overline{X}_1$ and $\overline{X}_2$. Such problems might arise in comparing a treated and a control population or weight gains of males and females, for instance. If a sample of size n_1 is taken from population 1 with mean μ_1 and variance σ_1^2, the distribution of the sample mean $\overline{X}_1$ will have expected value μ_1 and variance σ_1^2/n_1. Similarly, if from population 2 with mean μ_2 and variance σ_2^2, a sample of size n_2 is taken, the distribution of the sample mean $\overline{X}_2$ will have expected value μ_2 and variance σ_2^2/n_2. Now from section 4.5, the expected value of Y, the difference of the two means $(\overline{X}_1 - \overline{X}_2)$ will have the following mean and variance:

$$E(\overline{X}_1 - \overline{X}_2) = \mu_1 - \mu_2$$

$$Var(\overline{X}_1 - \overline{X}_2) = \frac{\sigma_1^2}{n_1} + \frac{\sigma_2^2}{n_2}$$

Further, if both original populations are normally distributed, $\overline{X}_1 - \overline{X}_2$ will be normally distributed. If the original populations are not normally distributed, then, by the central limit theorem, $\overline{X}_1 - \overline{X}_2$ will be approximately normally distributed for adequately large sample sizes.

6.2 t-Distribution

In example 6.2 we were able to calculate the probability that $3.41 < \overline{X} < 3.56$ by using the fact (discussed in section 5.4) that subtracting $\mu_{\overline{x}}$ from $\overline{X}$ and dividing by $\sigma_{\overline{x}}$ $(\sigma/\sqrt{n})$ reduced the distribution of $\overline{X}$ to the standard normal distribution. If σ (and hence $\sigma_{\overline{x}}$) is unknown and has to be replaced by the sample standard deviation s, subtracting $\mu_{\overline{x}}$ from $\overline{X}$ and dividing by $s/\sqrt{n}$ will not have a standard normal distribution. Instead it will have a *t-distribution,* which, although similar in shape to the standard normal distribution (bell-shaped), is described by a different mathematical formula and hence needs a different table for its evaluation. Further, there is not just a single t-distribution, but a family of them, each member of the family being distinguished by its value of n (or, as statisticians prefer, by its *degree of freedom,* denoted by f, where $f = n - 1$).

Whereas on the standard normal distribution $100(1 - \alpha)\%$ of the area lies between $\pm Z_{\alpha/2}$ (for instance, 95% lies between $\pm Z_{.025}$), on the t-distribution $100(1 - \alpha)\%$ lies between $\pm t_{\alpha/2}$ where $t_{\alpha/2}$ is defined for the t-distribution in the same way that $Z_{\alpha/2}$ was defined for the normal distribution. Thus 95% of the t-distribution lies between $\pm t_{.025}$ as shown in figure 6.1.

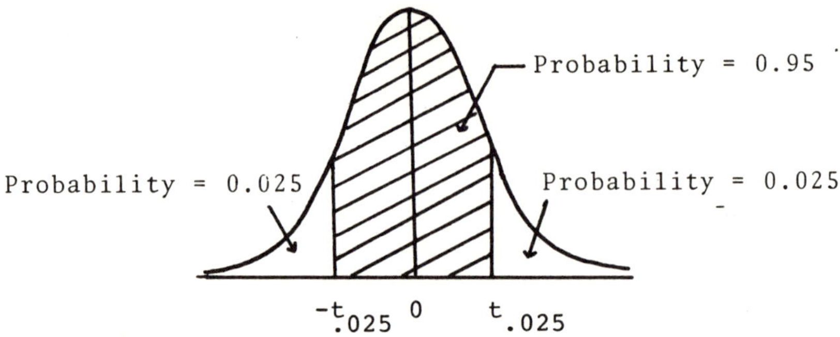

Fig. 6.1 *Central 0.95 probability of the t-distribution.*

The actual values of $t_{.025}$ change for different degrees of freedom and are given in Appendix Table 4. For other probabilities other values of t are required. For instance, 90% of the area lies between $\pm\ t_{.05}$ and 99% of the area lies between $\pm\ t_{.005}$.

Example 6.3 The reader should verify the following t values for various degrees of freedom using Appendix Table 4.

n	Degrees of freedom	Probability	t-Values
10	9	0.95	2.2622
14	13	0.90	1.7709
20	19	0.95	2.0930
30	29	0.99	2.7564
101	100	0.95	1.9840

One should note from Appendix Table 4 that as n becomes larger the t-distribution gets closer to the standard normal distribution (for instance, $t_{.025}$ gets closer to 1.96) and that, for all practical purposes, for $n > 30$, the two distributions are indistinguishable.

6.3 Chi-Square Distribution

Just as each of the $\binom{N}{n}$ possible samples has a mean $\overline{X}$, it also has a sample variance S^2. Calling the $\binom{N}{n}$ sample variances S_1^2, S_2^2, S_3^2, etc. the totality of these sample variances is called the *distribution of the sample variance.* Whereas $(\overline{X} - \mu_{\bar{x}})/s_{\bar{x}}$ has a t-distribution and

$(\overline{X} - \mu_{\bar{x}})/\sigma_{\bar{x}}$ has a standard normal distribution, the value $(n - 1)S^2/\sigma^2$ has what is called the *chi-square* (χ^2) *distribution.* To find the probability that $(n - 1)S^2/\sigma^2$ lies between two values a and b (with $a \leq b$), and hence to find the probability that S^2 lies between $a\sigma^2/(n - 1)$ and $b\sigma^2/(n - 1)$, one uses the tabulated probabilities of the χ^2-distribution in Appendix Table 5. As with the t-distribution, there is a family of χ^2-distributions, each one distinguished by its degrees of freedom. A typical χ^2-distribution is seen in figure 6.2. We should note its lack of symmetry, since $(n - 1)S^2/\sigma^2$ can never be negative. Whereas with the t-distribution 95% of the area lies between $-t_{.025}$ and $t_{.025}$, with the χ^2-distribution we have two values called $\chi^2_{.025}$ and $\chi^2_{.975}$ between which lies 95% of the χ^2-distribution as illustrated in figure 6.2. In general the area $100(1 - \alpha)\%$ lies between $\chi^2_{1-\alpha/2}$ and $\chi^2_{\alpha/2}$.

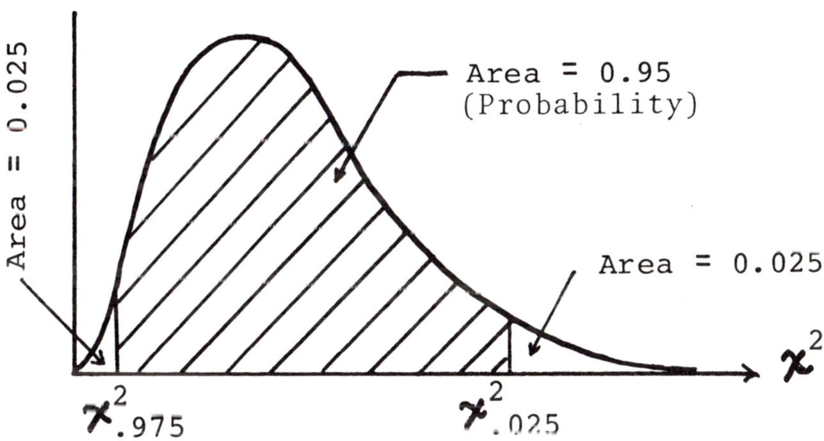

Fig. 6.2 *A typical* χ^2*-distribution.*

Example 6.4 Determining χ^2-values using Appendix Table 5.

n	f	Probability	χ^2-Values
10	9	0.95	$\chi^2_{.975} = 2.70$, $\chi^2_{.025} = 19.02$
20	19	0.95	$\chi^2_{.975} = 8.91$, $\chi^2_{.025} = 32.85$
15	14	0.90	$\chi^2_{.95} = 6.57$, $\chi^2_{.05} = 23.68$
26	25	0.99	$\chi^2_{.995} = 10.52$. $\chi^2_{.005} = 46.93$

6.4 Parameter Estimation and Confidence Intervals

In chapter 5, we always assumed that the parameters of the distributions (p, λ, μ, σ) were known and then went on to calculate various probabilities. We now consider the problem of using the information in a sample (the sample data or observations) to estimate the unknown parameters. This problem may be summarized as follows: We are dealing with a random variable x, the form of whose distribution is known, but it has unknown parameters. If x is continuous, we assume that it has a normal distribution; if it is discrete, it could have a binomial, Poisson, or hypergeometric distribution. We take a random sample and obtain the data $X3_1$, X_2, . . . , X_n. From these data we wish to estimate the parameter of interest using statistics calculated from the data. If we use a single statistic to estimate the parameter, we call the result a *point estimator.* A point estimator yields a single number with which to estimate the parameter. In general, point estimators are the "common sense" ones. For instance, in the normal distribution we use $\overline{X}$, the sample mean, as an estimator of μ, the population mean; we use S^2, the sample variance, as an estimator of σ^2, the population variance; in the binomial distribution we use X/n, the sample proportion of defectives, to estimate p, the population proportion of defectives. A particular value of an estimator for a particular set of data is called an *estimate.* While $\overline{X}$ might be the estimator of μ, for a particular experiment we might have $\overline{X} = 15.7$ as an estimate of μ. Since the particular sample is simply one among the $\binom{N}{n}$ possible ones, and its statistic one among the $\binom{N}{n}$ possible ones, point estimates give no indication of how precise they are.

To obtain some idea of how good the estimates are, we calculate an interval that has a known probability of containing the unknown parameter. We call such an interval, with its prescribed probability containing the specified parameter within it, a *confidence interval.* Estimators of this type are called *interval estimators,* and their particular values for a specific set of data are called *interval estimates.* A short confidence interval indicates a precise estimate and a long confidence interval an imprecise one. For this reason, interval estimators are generally more desirable and useful. We shall subsequently discuss confidence intervals for estimating μ, σ^2, and σ of the normal distribution and p of the binomial distribution.

6.4.1 Confidence Intervals for the Mean of a Normal Distribution

If we wish to estimate the mean μ of a normal distribution when the variance σ^2 is known, we know from chapter 5 that

$$z = \frac{\overline{X} - \mu}{\sigma/\sqrt{n}}$$

has a standard normal distribution and that the only unknown quantity is μ. Now, we can write

$$P[-1.96 < z < 1.96] = 0.95$$

which, after replacing z with its equivalent, gives

$$P\left[-1.96 < \frac{\overline{X} - \mu}{\sigma\sqrt{n}} < 1.96\right] = 0.95$$

But this (after some manipulation) is the same as

$$P\left[\overline{X} - 1.96\frac{\sigma}{\sqrt{n}} < \mu < \overline{X} + 1.96\frac{\sigma}{\sqrt{n}}\right] = 0.95$$

The expression inside the square brackets constitutes an interval with two end points, the lower being $\overline{X} - 1.96\ \sigma/\sqrt{n}$ and the upper being $\overline{X} + 1.96\ \sigma/\sqrt{n}$, which contains μ. The probability that this interval contains μ is 0.95, and, hence, the interval is called a 95% confidence interval for μ. The general $100(1 - \alpha)\%$ level of confidence interval is

$$\left(\overline{X} - Z_{\alpha/2}\frac{\sigma}{\sqrt{n}}, \overline{X} + Z_{\alpha/2}\frac{\sigma}{\sqrt{n}}\right)$$

Since $\dfrac{\overline{X} - \mu}{\sigma/\sqrt{n}}$ is (by the central limit theorem) approximately normal for large sample sizes even when the original distribution is nonnormal, the above confidence interval gives an approximate confidence interval for the means of nonnormal populations, provided that the sample size is adequately large (if $n > 30$).

Example 6.5 In a study of milk fat content, it was known from extensive past records that $\sigma = 0.3$. A particular milk producer has the following random sample of milk fat contents recorded: 3.0, 3.2, 2.8, 3.4, 3.1. Calculate a 95% confidence interval for his population mean μ for milk fat content. Here $\overline{X} = 3.1$, $\sigma = 0.3$, so that a 95% confidence interval is

$$\left(\overline{X} - 1.96\frac{\sigma}{\sqrt{n}}, \overline{X} + 1.96\frac{\sigma}{\sqrt{n}}\right)$$

or

$$\left(\overline{X} - \frac{1.96(0.3)}{\sqrt{5}}, \overline{X} + \frac{1.96(0.3)}{\sqrt{5}}\right)$$

or between 2.84 and 3.36. From these results we can state that the probability is 0.95 that the population mean of milk fat content lies between 2.84 and 3.36. We can state that we are 95% confident that μ lies between 2.84 and 3.36. If, as is often the case, the variance σ^2 is unknown, the above confidence intervals cannot be calculated and therefore are useless. However, if we remember that

$$t = \frac{\overline{X} - \mu}{S/\sqrt{n}}$$

has a t-distribution, and

$$P\left[-t_{.025} < \frac{\overline{X} - \mu}{S/\sqrt{n}} < t_{.025}\right] = 0.95$$

or

$$P\left[\overline{X} - t_{.025}\frac{S}{\sqrt{n}} < \mu < \overline{X} + t_{.025}\frac{S}{\sqrt{n}}\right] = 0.95$$

we get a 95% confidence interval for μ when σ^2 is unknown. Other levels of confidence can be similarly considered (see exercise 5).

We have seen that if the population variance σ^2 is unknown, we must resort to using the t-distribution. But we know from section 6.1 for large n that the difference between the t-distribution and the standard normal distribution is negligible. Hence, if σ^2 is unknown, but n is large ($n > 30$), we may use the standard normal distribution. We sometimes refer to the confidence intervals using the standard normal distribution as large-sample estimators and those based on the t-distribution as small-sample estimators.

Example 6.6 To estimate the mean length μ of a population of cucumbers which are assumed to be normally distributed, a random sample of 20 was selected for which $\overline{X}$ and S were calculated to be 0.4 m and 0.1 m, respectively. A 90% confidence interval for μ is

$$\left(\overline{X} - t_{.05}\frac{S}{\sqrt{n}}, \overline{X} + t_{.05}\frac{S}{\sqrt{n}}\right)$$

or
$$\left(0.4 - \frac{(1.73)(0.1)}{\sqrt{20}}, 0.4 + \frac{(1.73)(0.1)}{\sqrt{20}}\right)$$

or
$$(0.31, 0.49)$$

In summary, the results for calculating a confidence interval for the population mean μ are as follows:

	Original distribution			
	Normal		Nonnormal	
	n small	n large	n small	n large
known σ^2	z-table	z-table	not applicable	z-table
unknown σ^2	t-table	z-table	not applicable	z-table

6.4.2 Confidence Interval for the Variance and Standard Deviation of a Normal Distribution

As we saw in section 6.4.1, sometimes the variance σ^2 of a normal distribution is unknown and needs to be estimated. As mentioned in section 6.4, S^2 is a point estimator of σ^2. In this section we consider an interval estimation of σ^2.

Since $(n - 1)S^2/\sigma^2$ has a χ^2-distribution, we can find the two numbers $\chi^2_{1-\alpha/2}$ and $\chi^2_{\alpha/2}$ which have a prescribed probability of containing $(n - 1)S^2/\sigma^2$. In particular, the probability is 0.95 that a χ^2-distribution lies between $\chi^2_{.975}$ and $\chi^2_{.025}$, thus

$$P\left[\chi^2_{.975} < \frac{(n-1)S^2}{\sigma^2} < \chi^2_{.025}\right] = 0.95$$

which, after some manipulation, yields

$$P\left[\frac{(n-1)S^2}{\chi^2_{.025}} < \sigma^2 < \frac{(n-1)S^2}{\chi^2_{.975}}\right] = 0.95$$

The expression inside the square brackets is a 95%confidence interval for σ^2. More generally, a $100(1 - \alpha)\%$ confidence interval for σ^2 is

$$\left(\frac{(n-1)S^2}{\chi^2_{\alpha/2}}, \frac{(n-1)S^2}{\chi^2_{1-\alpha/2}}\right)$$

A confidence interval for σ is found by taking the square root of that for σ^2. A $100(1 - \alpha)\%$ confidence interval for σ is

$$\sqrt{\frac{(n-1)S^2}{\chi^2_{\alpha/2}}}, \quad \sqrt{\frac{(n-1)S^2}{\chi^2_{1-\alpha/2}}}$$

Example 6.7 Continuing with example 6.6, suppose that we wish to calculate 95% confidence intervals for σ^2 and σ, the variance and standard deviation of cucumber lengths. We take a random sample of 20 cucumbers and calculate $S^2 = 0.01$ m^2 and $S = 0.1$ m.

Hence $n = 20$, degrees of freedom $= 19$, $\alpha = 0.05$, $\chi^2_{.975} = 8.91$, $\chi^2_{0.025} = 32.85$. The 95% confidence interval for σ^2 is

$$\left(\frac{(19)(0.01)}{32.85}, \frac{(19)(0.01)}{8.91}\right)$$

or (0.0058, 0.0213). Similarly, the 95% confidence interval for σ is (0.076, 0.146).

6.4.3 Confidence Interval for a Proportion

If a discrete random variable has a binomial distribution, we might be concerned with estimating the population proportion of defectives, p. We take a random sample of size n and obtain X defectives, giving X/n as the proportion of defectives in the sample. If n is small, graphical solutions for 95% and 99% confidence intervals are given in Appendix Tables 6.1 and 6.2. These tables can also be used for large n, but are less accurate because of interpolation problems.

Example 6.8 From a large consignment of apples, a random sample of 100 apples yields 15 bad ones. Calculate a 95% confidence interval for p, the proportion of bad ones in the whole consignment.

Here $n = 100$, $X = 15$, $X/n = 0.15$. From table 6.1 we read $X/n = 0.15$ along the lower horizontal axis and then go up vertically until we reach the two contours labeled "100". The numbers on the contour lines correspond to n. The confidence limits are read off on the left-hand scale. Thus, a 95% confidence interval is (0.08, 0.23). If X/n is greater than 0.5, we read it on the upper horizontal scale and read the confidence interval on the right-hand scale.

When n is large and p is not too close to 1 or 0, the central limit theorem allows us to use the normal approximation to the binomial, since, from section 5.2, if x is binomial, it has expected value np and variance $np(1 - p)$. From section 5.2, X/n has expected value p and variance $p(1 - p)/n$. From the central limit theorem

$$\frac{X/n - p}{\sqrt{\dfrac{p(1 - p)}{n}}}$$

has approximately a standard normal distribution. If we wish to calculate a 95% confidence interval, then

$$P\left[-1.96 < \frac{X/n - p}{\sqrt{\dfrac{p(1 - p)}{n}}} < 1.96\right] = 0.95$$

and a 95% confidence interval for p is

$$\left(\frac{X}{n} - 1.96\sqrt{\frac{p(1 - p)}{n}}, \frac{X}{n} + 1.96\sqrt{\frac{p(1 - p)}{n}}\right)$$

This interval depends on p, which is, of course, unknown; however, if we replace p by X/n, we obtain the following approximate

95% confidence interval for p:

$$\left(\frac{X}{n} - 1.96\sqrt{\frac{\frac{X}{n}\left(1-\frac{X}{n}\right)}{n}},\ \frac{X}{n} + 1.96\sqrt{\frac{\frac{X}{n}\left(1-\frac{X}{n}\right)}{n}}\right)$$

Example 6.9 Reconsider example 6.8 where $n = 100$, $X = 15$, $X/n = 0.15$, $\sqrt{\frac{\frac{X}{n}\left(1-\frac{X}{n}\right)}{n}} = 0.0357$. A 95% confidence interval for p is (0.09, 0.23).

6.4.4 Confidence Interval for the Difference of Two Means

Some problems ask the experimenter to compare the difference of two population means. For example, when comparing a treated group against its control, or comparing two geographical regions, we are dealing with two different populations, each having a mean. If our two populations, labeled 1 and 2, have respective means μ_1 and μ_2, our objective is to estimate $\mu_1 - \mu_2$.

First, we consider the case where both population variances are assumed to be known (perhaps from previous similar experiments). From population 1, having mean μ_1 and variance σ_1^2, a sample of n_1 observations $X_{11}, X_{12}, \ldots, X_{1n}$ is selected and the sample mean calculated to be $\overline{X}_1$. From population 2, having mean μ_2 and variance σ_2^2, a sample of n_2 observations is selected independently of the first sample, and the sample mean calculated to be $\overline{X}_2$. A point estimator of $\mu_1 - \mu_2$ is simply $\overline{X}_1 - \overline{X}_2$. We will be more concerned with an interval estimator of $\mu_1 - \mu_2$.

From the section 6.1, if the original populations are normal, $\overline{X}_1 \sim N(\mu_1, \sigma_1^2/n_1)$, and $\overline{X}_2 \sim N(\mu_2, \sigma_2^2/n_2)$, and $\overline{X}_1 - \overline{X}_2 \sim N(\mu_1 - \mu_2, \sigma_1^2/n_1 + \sigma_2^2/n_2)$. If the original populations are not normally distributed, $\overline{X}_1 - \overline{X}_2$ will be approximately normally distributed.

Thus, $\dfrac{\overline{X}_1 - \overline{X}_2 - (\mu_1 - \mu_2)}{\sqrt{\dfrac{\sigma_1^2}{n_1} + \dfrac{\sigma_2^2}{n_2}}}$ has a standard normal distribution, and

$$P\left[-1.96 < \frac{\overline{X}_1 - \overline{X}_2 - (\mu_1 - \mu_2)}{\sqrt{\frac{\sigma_1^2}{n_1} + \frac{\sigma_2^2}{n_2}}} < 1.96\right] = 0.95$$

yields the following 95% confidence interval for $\mu_1 - \mu_2$:

$$\left(\overline{X}_1 - \overline{X}_2 - 1.96\sqrt{\frac{\sigma_1^2}{n_1} + \frac{\sigma_2^2}{n_2}}, \overline{X}_1 - \overline{X}_2 + 1.96\sqrt{\frac{\sigma_1^2}{n_1} + \frac{\sigma_2^2}{n_2}}\right)$$

Example 6.10 The weight at 20 weeks of two populations of broiler chickens is to be compared. Group 1 has been fed a weight-increasing hormone and group 2 has been fed the standard diet. The weights of both groups are assumed to be normally distributed, and from previous similar studies the variances are known to be $\sigma_1^2 = 0.8$ kg^2 and $\sigma_2^2 = 1.2$ kg^2. From group 1 a sample of 20 yields $\overline{X}_1 = 6.3$ kg and 30 observations from group 2 yield $\overline{X}_2 =$ 5.7 kg. Calculate a 90% confidence interval for the difference $\mu_1 - \mu_2$ of the population means.

Here $n_1 = 20$, $n_2 = 30$, $\sigma_1^2 = 0.8$, $\sigma_2^2 = 1.2$, $\overline{X}_1 = 6.3$, $\overline{X}_2 =$ 5.7. The 90% confidence interval is

$$(0.6 - (1.65)\sqrt{0.08}, 0.6 + (1.65)\sqrt{0.08})$$

or (0.133, 1.067).

If the two variances are unknown but may be presumed to be equal, we then proceed as follows: (We shall not consider what to do if the variances are unknown and unequal, but the interested reader may find this discussed in Malik and Mullen 1971). From the two samples, we calculate the sample variances S_1^2 and S_2^2 by the usual method. Since both of these sample variances are estimates of the common population variance, which we call σ^2, we must construe them to obtain a *pooled variance estimator.* Calling the pooled variance estimator S^2, we have

$$S^2 = \frac{(n_1 - 1)S_1^2 + (n_2 - 1)S_2^2}{n_1 + n_2 - 2}$$

Note that this method of pooling S_1^2 and S_2^2 reflects the sample sizes n_1 and n_2 and gives more weight to the larger sample. Now, since σ_1^2 and σ_2^2 are both replaced by σ^2, which, in turn, is replaced by S^2, we may use the t-table with $n_1 + n_2 - 2$ degrees of freedom in the manner of 6.4.1 to give

$$P\left[-t_{.025} < \frac{\overline{X}_1 - \overline{X}_2 - (\mu_1 - \mu_2)}{S\sqrt{\dfrac{1}{n_1} + \dfrac{1}{n_2}}} < t_{.025}\right] = 0.95$$

which, when rearranged, yields

$$\left(\overline{X}_1 - \overline{X}_2 - t_{.025}\, S\sqrt{\frac{1}{n_1} + \frac{1}{n_2}},\ X_1 - X_2 + t_{.025}\, S\sqrt{\frac{1}{n_1} + \frac{1}{n_2}}\right)$$

as a 95% confidence interval for $\mu_1 - \mu_2$.

Example 6.11 Suppose that in example 6.10 the variances σ_1^2 and σ_2^2 are unknown but presumed to be equal. The sample of 20 observations from population 1 yields $\overline{X}_1 = 6.3$ and $S_1^2 = 1.0$, while the sample of 30 observations from population 2 yields $\overline{X}_2 = 5.7$ and $S_2^2 = 1.5$. Calculate a 95% confidence interval for $\mu_1 - \mu_2$.

The value of S^2, the pooled variance estimator, is

$$S^2 = \frac{(19)(1.0) + (29)(1.5)}{48} = 1.302$$

so that $S = 1.141$. The 95% confidence interval for $\mu_1 - \mu_2$ is

$$\left(0.6 - (1.96)(1.141)\sqrt{\frac{1}{20} + \frac{1}{30}},\ 0.6 + (1.96)(1.141)\sqrt{\frac{1}{20} + \frac{1}{30}}\right)$$

Notice that since we need to use the t-table with 48 degrees of freedom, we may replace the value of $t_{.025}$ by $Z_{.025}$.

In some experimental situations concerning two populations the data occur in pairs. For instance, before treatment versus after treatment of the same subject or object, indoor response versus outdoor response, or procedure 1 versus procedure 2 on the same object. The data may be represented as follows:

Pair	Treatment 1	Treatment 2
1	X_{11}	X_{12}
2	X_{21}	X_{22}
3	X_{31}	X_{32}
.	.	.
.	.	.
.	.	.
n	X_{n1}	X_{n2}

where the n pairs (on n subjects) are regarded as being independent. Since for each pair the two observations are on the same subject or object, the data from treatments 1 and 2 cannot be regarded as being independent, and hence the previous results of this section do not apply.

If treatment 1 and treatment 2 both have normal distributions with means μ_1 and μ_2, the differences for each pair $(X_{1i} - X_{2i})$, which we will call D_i, will also be normally distributed with mean $\mu_1 - \mu_2$. If the D_i's have a normal distribution with mean $\mu_1 - \mu_2$ and unknown variance σ^2 which can be estimated by

$$S_D^2 = \frac{\Sigma(D_i - \overline{D})^2}{n - 1}$$

where $\overline{D}$ is the mean of the D_i's, $\overline{D}$ will also have a normal distribution with mean $\mu_1 - \mu_2$ and variance σ^2/n. Therefore,

$$\frac{\overline{D} - (\mu_1 - \mu_2)}{S_D/\sqrt{n}}$$

will have a t-distribution with n − 1 degrees of freedom. Thus, a 95% confidence interval for $\mu_1 - \mu_2$ is

$$\left(\overline{D} - t_{.025}\frac{S_D}{\sqrt{n}}, \overline{D} + t_{.025}\frac{S_D}{\sqrt{n}}\right)$$

Example 6.12 To compare the results of two laboratories' measurement of calcium levels in food, 10 different food products were split in 2 parts, part 1 and part 2. The part 1 samples were sent to one laboratory and the part 2 samples to a second to be analyzed for their calcium levels. For the results (in percentages) that follow, calculate a 95% confidence interval for the difference of the two laboratories.

Laboratory 1	Laboratory 2	Difference
10.7	9.8	0.9
12.3	12.7	−0.4
12.1	12.8	−0.7
9.0	8.6	0.4
14.2	13.0	1.2
9.8	9.3	0.5
5.6	6.5	−0.9
8.7	8.1	0.6
13.6	13.9	−0.6
12.9	11.5	1.4

We have $\overline{D} = 0.24$, $S_D = 0.83$. A 95% confidence for $\mu_1 - \mu_2$ is

$$\left(\overline{D} - t_{.025}\frac{S_D}{\sqrt{n}}, \overline{D} + t_{.025}\frac{S_D}{\sqrt{n}}\right)$$

or (−0.35, 0.83).

Exercises

1. Use the data of example 6.1 to take all possible samples of size 3, calculate the mean of each, and show that $\mu_{\overline{X}} = \mu$ and that $\sigma^2_{\overline{X}} = \sigma^2(N - n)/n(N - 1)$.

2. If the weight of a population of sacks of potatoes is assumed to be normally distributed, with mean $\mu = 100$ lbs and variance $\sigma^2 = 25$, what is the probability that the mean weight of N sacks will be less than 98 lbs for N = 1, 9, 16, 25?

3. If the weights of cans of peaches are normally distributed, with $\mu = 1$ kg and $\sigma^2 = 0.25$, what is the probability that the mean weight of a sample of 20 cans is
   (a) between 0.9 and 1.1
   (b) between 0.98 and 1.2
   (c) less than 0.98?

4. If the volume of bottles of ketchup is assumed to be normally distributed, with unknown mean μ but known variance $\sigma^2 = 25\ cc^2$ and a sample of size 20 yields a mean volume of 200 cc, calculate (i) a 95%, and (ii) a 99% confidence interval for μ.

5. In exercise 4 if the variance is unknown and, from the sample, we calculate $S^2 = 30$, calculate 95% and 99% confidence intervals for μ.

6. If the shelf life of a particular dairy product is assumed to be normally distributed, with standard deviation $\sigma = 2$ days and an experiment with $n = 15$ yields $\overline{X} = 14$ days, calculate a 90% confidence interval for μ, the population mean life length.

7. Rework exercise 6 if σ is unknown but the sample S is calculated to be 3 days.

8. Show that the length of a confidence interval for μ (with σ^2 unknown) is $2\sigma Z_{\alpha/2}/\sqrt{n}$.

9. Use the results of exercise 8 to calculate how large n must be in exercise 4 to reduce the length of the 95% confidence interval to 2 cc.

10. In exercise 5 calculate 90% and 95% confidence intervals for σ^2 and σ.

11. In exercise 7 calculate 95% confidence intervals for σ^2 and σ.

12. In examining boxes of incoming fruit, the quality manager judges a box to be either ripe or unripe. Of 50 random boxes examined, 20 were found to be unripe. Calculate 95% and 99% confidence intervals for p, the population proportion of unripe boxes, using Appendix Tables 6 and 7.

13. Recalculate exercise 12 using the normal approximation technique of section 6.4.3.

14. To compare the results of two treatments on the yield of corn, 25 plots are given treatment 1 and 20 other plots are

given treatment 2. The variances are known from prior experiments to be $\sigma_1^2 = 5.3$ and $\sigma_2^2 = 8.4$. If $\overline{X}_1 = 83$ and $\overline{X}_2 = 64$, calculate (i) a 90%, and (ii) a 95% confidence interval for $\mu_1 - \mu_2$, the difference in yield of the two treatments, assuming that yield has a normal distribution.

15. Suppose that the variances in exercise 14 are unknown, and that from the samples we calculate that $S_1^2 = 6.4$ and $S_2^2 = 7.8$. Assuming that the variances are equal, calculate a 90% confidence interval for $\mu_1 - \mu_2$.

16. The inner and outer diameters of washers are assumed to have normal distributions with means μ_1 and μ_2, respectively, and unknown variances. The thickness of the washers is the difference between the outer and inner diameters. To estimate the mean thickness of the washers, a random sample of 8 washers yielded the following results:

Outer diameter	Inner diameter
.24	.19
.25	.18
.22	.18
.24	.19
.23	.16
.20	.16
.28	.21
.26	.23

Calculate a 90% and a 95% confidence interval for the mean thickness of the washers.

7. Hypothesis Testing

Whereas statistical estimation uses sample observations to form point or interval estimates of unknown parameters, a second major area of statistics uses observations to test whether or not certain assumptions about the parameters are true. These assumptions are referred to as *hypotheses* and this chapter is concerned with techniques for testing the validity of these hypotheses.

7.1 Null and Alternate Hypotheses

Consider a problem where regulations require that the mean weight of a large consignment of sacks of potatoes be at least 20 kg. To see if this is so, we select n sacks of potatoes, weigh them, and obtain as observations the weights $X_1, X_2, \ldots, X_n$. Then, if μ represents the mean weight of sacks of potatoes, we are inquiring if $\mu \geq 20$; alternatively, if the regulations are not satisfied, $\mu < 20$.

More generally we will be dealing with a random variable whose distribution is assumed to be known but whose parameters are unknown. Unlike problems in estimation where we wish to estimate the value of the parameter, here we wish to decide if the parameter is equal to, greater than, or less than a preassigned or presumed value. Problems of this type are called *hypothesis testing* problems.

Example 7.1 We might hypothesize that the average weight of a particular type of imported orange does not exceed 200 gm and test this by taking a random sample of 500 oranges from an incoming consignment.

Example 7.2 In trying to control the weight of cans of peaches on a production line, we might hypothesize that the median weight of a can is greater than or equal to 32 oz.

In example 7.1 we talk of testing the hypothesis that μ, the mean weight of this type of orange, is less than or equal to 200 gm. As we shall see, there may be several hypotheses to be tested. The main one is usually referred to as the *null hypothesis,* denoted by H_0. In example 7.1 we write H_0: $\mu \leq 200$ and in example 7.2 we write H_0: $\mu \geq 32$. Any other hypotheses are referred to as *alternate hypotheses,* denoted by H_A. For examples 7.1 and 7.2 the alternate hypotheses are

$$H_A\colon \mu > 200 \quad \text{and} \quad H_A\colon \mu < 32$$

respectively.

7.2 Type I and Type II Errors

Hypotheses are tested on the basis of a random sample. We conclude either that the null hypothesis is true (H_0 is accepted) or that the null hypothesis is false (H_0 is rejected and H_A is accepted). Because of the lack of certainty inherent in using a random sample to draw conclusions about a parameter value, it is possible to make errors. For example, in the illustration of the weights of sacks of potatoes, if, in fact, the mean weight is greater than 20 and yet, because of sampling variation, it is concluded that $\mu < 20$, we have made the error of rejecting H_0 when H_0 is true. Rejecting H_0 when H_0 is true is a *type I error.* If we conclude that $\mu \geq 20$ when, in fact, $\mu < 20$, we wrongly accept H_0. Accepting H_0 when H_0 is false is a *type II error.* Of course, we will not know if an error is made because the value of the parameter is unknown. Such errors should not be regarded as blunders, but as an inherent part of statistical testing. Since it is impossible not to make some errors, we hope that the probability of such errors is small. The probability of a type I error is denoted by α and referred to as the *significance level* of the test. The probability of type II error is denoted by β. The probability of correctly rejecting H_0 is thus $1 - \beta$ and referred to as the *power* of the test. The following illustrates the two decisions possible and the associated errors.

True situation	Decision	
H_0 True	Accept H_0 No error	Reject H_0 Type I error (Probability α)
H_0 False	Type II error (Probability β)	No error

It is customary to prespecify the value of α desired, (pragmatically acceptable values are usually 01., 0.05, and 0.01 although any other values may be chosen) and to control the value of β by varying the sample size. Except in simpler situations, the calculation of β (for a chosen value of α) is complicated and rarely done. Hence, in most testing situations the probability of making a type I error is known but the probability of making a type II error is usually unknown. If we reject the null hypothesis, the probability of error is known, whereas if we accept the null hypothesis, the probability of error is unknown. For this reason, rather than say "accept the null hypothesis," we shall follow the convention of many authors and say, "We cannot reject the null hypothesis." This second statement is weaker than the first and leaves open the possibility that future studies with more data might reject the null hypothesis.

As a consequence of not using the expression "We accept the null hypothesis," we usually regard the null hypothesis as the status quo hypothesis and the alternative hypothesis as the one reflecting any change due to the experiment. If a standard diet produces chicken eggs with a mean weight of 55 gm or less and a proposed new diet is hypothesized to increase the mean egg weight, we would state

$$H_0: \mu \leq 55, \qquad H_A: \mu > 55$$

Reconsider the discussion concerning the weights of sacks of potatoes and suppose that the null and alternate hypotheses are $\mu \geq 20$ and $\mu < 20$, respectively, or

$$H_0: \mu \geq 20, \qquad H_A: \mu < 20$$

Further, for simplicity, assume that the random variable "weight of a sack of potatoes" has a normal distribution whose standard deviation (known perhaps from previous similar studies) is $\sigma = 20$. Let us suppose that we are willing to accept a value of α, the probability of a type I error, of 0.05 and that we perform the test by taking the weights of a random sample of 25 sacks. We know that $\overline{X}$, the mean weight of the 25 observations, has a normal distribution with standard deviation $20/\sqrt{25}$ or 0.4. If the null hypothesis is true (if $\mu \geq 20$), the probability that $\overline{X}$ is less than $20 - (1.65)(.4)$, or 19.34, is 0.05 or less (we know that 5% of the normal distribution lies below $\mu - 1.96\sigma/\sqrt{n}$). If we reject H_0 when $\overline{X} < 19.34$, we will be performing the tests with an α level of 0.05 or less.

Note our reasoning process: we substituted $\mu = 20$ in place of the null hypothesis H_0: $\mu \geq 20$. For this reason we, along with many authors, follow the convention of writing H_0: $\mu = 20$ in place of the null hypothesis. Next we calculate $20 - 1.65\sigma/\sqrt{n} = 19.34$ and reject H_0 if $\overline{X} < 19.34$. In general, to test

$$H_0: \mu \geq \mu_0 \text{ against } H_A: \mu < \mu_0$$

with an α level of 0.05, we reject H_0 if $\overline{X} < \mu_0 - 1.65\sigma/\sqrt{n}$. For a general level of α, we reject H_0 for $\overline{X} < \mu_0 - Z_\alpha\sigma/\sqrt{n}$. Alternatively, we could reject H_0 if

$$\frac{\overline{X} - \mu_0}{\sigma/\sqrt{n}} < -Z_\alpha$$

and in fact, this is the usual way of performing the test. The expression $(\overline{X} - \mu_0) \div \sigma/\sqrt{n}$ is called the *test statistic* and labeled z (since the test statistic has a standard normal distribution). Values of the test statistic less than $-Z_\alpha$ are called the *critical region* and values greater than $-Z_\alpha$ are called the *acceptance region.* If the test statistic falls in the critical region, the null hypothesis is rejected.

The null hypothesis is not rejected if z falls in the acceptance region.

To calculate the value of β, the probability of a type II error, assume that the value of μ is 19 (actually unknown to us). Since a type II error is committed if H_0 is wrongly accepted, we wish to know the probability of accepting H_0 (this is done if $\overline{X} > 19.34$) when $\mu = 19$. But $\overline{X} > 19.34$ is equivalent to

$$\frac{\overline{X} - 19}{2/\sqrt{25}} > \frac{19.34 - 19}{2/\sqrt{25}} = 0.85$$

or $z > 0.85$. From Appendix Table 3, the probability that $z > 0.85$ is 0.1977. If μ had been 19.5, β would have been 0.6554. Other values of μ and β are given as follows:

μ	β
20	0.9500
19.5	0.6554
19.0	0.1977
18.5	0.0179
18.0	0.0004

7.3 Types of Tests

7.3.1 One- and Two-Tailed Tests

In the discussion of testing the weights of sacks of potatoes, the alternative hypothesis was H_A: $\mu < 20$, and we rejected H_0 if the test statistic z was too small (smaller than $-Z_\alpha$). Such a test is called a *one-sided* or *one-tailed test.* Similarly, the test H_0: $\mu \leq \mu_0$ against H_A: $\mu > \mu_0$ is one-tailed. We calculate the test statistic z and reject the null hypothesis if z is too large, that is, if $z > Z_\alpha$. Frequently we wish to reject the null hypothesis if μ is larger or smaller than μ_0. We usually write H_0: $\mu = \mu_0$ against H_A: $\mu \neq \mu_0$, so that the alternative hypothesis includes both $\mu < \mu_0$ and $\mu > \mu_0$. Such a test is called a *two-sided* or *two-tailed test.* We reject the null hypothesis if the test statistic z is too large or too small. To maintain the probability of a type I error being α, we reject H_0 if $z > Z_{\alpha/2}$ or if $z < -Z_{\alpha/2}$. These one- and two-sided tests are illustrated in figure 7.1 where the curves represent standard normal distributions.

It may happen that a test statistic is significant and the null hypothesis is rejected if a two-sided test is used but not significant if a one-sided test is used. Generally an experimenter must choose before obtaining experimental data whether a test is one- or two-sided. The nature of the experiment generally dictates which is appropriate. It is certainly unscientific to run one- and two-sided tests with the same data and pick the result best liked.

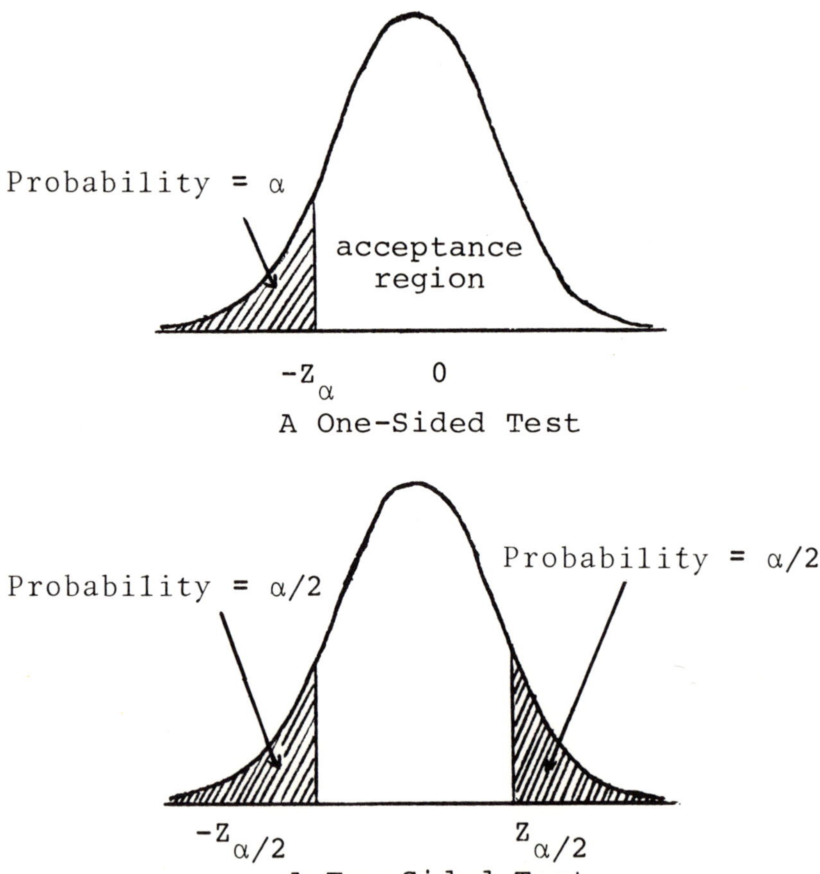

Fig. 7.1 *Acceptance and critical regions for one- and two-sided tests.*

7.3.2 Testing the Mean of a Normal Distribution

Z-Test

Suppose that we are dealing with a random variable of normal distribution whose variance σ^2 is known. We wish to test the following null and alternative hypotheses:

$$H_0: \mu \geq \mu_0, \qquad H_A: \mu < \mu_0$$

where μ_0 is a number specified by the experimenter. We are willing to accept the probability of a type I error of α and take a random sample of size n, $X_1, X_2, \ldots, X_n$, calculate $\overline{X}$, and from it calculate the test statistic

$$z = \frac{\overline{X} - \mu_0}{\sigma/\sqrt{n}}$$

From section 7.2 we reject the null hypothesis if z is too small, that is, smaller than $-Z_\alpha$.

If we wish to test

$$H_0: \mu \leq \mu_0 \qquad H_A: \mu > \mu_0$$

the critical region is $z > Z_\alpha$. For the two-sided test

$$H_0: \mu = \mu_0, \qquad H_A: \mu \neq \mu_0$$

the critical region is $z < -Z_{\alpha/2}$ or $z > Z_{\alpha/2}$.

Example 7.3 It is hypothesized that the mean weight of chicken eggs produced by hens fed a particular diet is higher than 55 gm. An experimenter wishing to test this assumes that egg weight has a normal distribution (from past experience $\sigma = 4$) and decides to test

$$H_0: \mu = 55, \qquad H_A: \mu > 55$$

using a one-sided test with significance level of 5%. The experiment takes a random sample of 25 egg weights and calculates $\overline{X} = 56.0$. Is there enough evidence to reject the null hypothesis?

The test statistic is

$$z = \frac{\overline{X} - \mu_0}{\sigma/\sqrt{n}} = \frac{56.0 - 55}{4/5} = 1.25$$

Since we reject H_0 if $z > Z_{.05} = 1.65$ (Appendix Table 3), we conclude that the null hypothesis cannot be rejected in this case. Without further measuring, we have no evidence to conclude that the new diet increases average egg weight.

t-Test
If the random variable under consideration has a normal distribution whose variance σ^2 is unknown, the test statistic z cannot be calculated (since it depends on σ). In this case we calculate

$$t = \frac{\overline{X} - \mu_0}{S/\sqrt{n}}$$

and compare it to the t-distribution with $n - 1$ degrees of freedom. To test H_0: $\mu \leq \mu_0$ against H_A: $\mu > \mu_0$ with level of significance α, reject H_0 if $t > t_\alpha$. Similarly, for H_0: $\mu \geq \mu_0$ against H_A: $\mu < \mu_0$, reject H_0 if $t < -t_\alpha$. For the two-sided test reject H_0 if $t < -t_{\alpha/2}$ or if $t > t_{\alpha/2}$.

Example 7.4 Reconsider example 7.3 but suppose that σ is unknown. The null and alternate hypotheses are the same, namely, H_0: $\mu = 55$, H_A: $\mu > 55$, and α is still 0.05. From the random sample of 25 observations from the normal distribution of egg weights, suppose that $\overline{X} = 56.0$ and $S = 6.0$. The test statistic is

$$t = \frac{\overline{X} - \mu_0}{S/\sqrt{n}} = \frac{56.0 - 55.0}{6/5} = 0.83$$

From Appendix Table 4 the value of t_α for $n - 1 = 25 - 1$ degrees of freedom is 1.71. We cannot reject the null hypothesis. We have no evidence to conclude that the new diet increases mean egg weight.

7.4 Comparison of Hypothesis Testing and Confidence Intervals

We can surmise, at this point, that there is a relationship between hypothesis testing and confidence intervals. This relationship is more apparent if we consider the two-sided test. For a significance level α and known σ we have

$$H_0: \mu = \mu_0, \qquad H_A: \mu \neq \mu_0,$$

for which the critical region is $z < -Z_{\alpha/2}$ or $z > Z_{\alpha/2}$ and the acceptance region is $-Z_{\alpha/2} < z < Z_{\alpha/2}$. From section 6.4.1, a $100(1 - \alpha)\%$ confidence interval for μ (which has probability $1 - \alpha$) is as follows:

$$P\left[\overline{X} - Z_{\alpha/2}\frac{\sigma}{\sqrt{n}} < \mu < \overline{X} + Z_{\alpha/2}\frac{\sigma}{\sqrt{n}}\right] = 1 - \alpha$$

which is equivalent to

$$P\left[-Z_{\alpha/2} < \frac{\overline{X} - \mu}{\sigma/\sqrt{n}} < Z_{\alpha/2}\right] = 1 - \alpha$$

But $\dfrac{\overline{X} - \mu}{\sigma/\sqrt{n}} = z$, which shows that a $100(1 - \alpha)\%$ confidence interval for μ is the same as the acceptance region in a two-sided test with significance level α. In other words, accepting H_0 at significance level α in a two-sided test, is equivalent to μ_0 being contained in $100(1 - \alpha)\%$ confidence interval. Conversely, rejecting H_0 at significance level α in a two-sided test is equivalent to μ_0 not being contained in the $100(1 - \alpha)\%$ confidence interval.

Similar results hold for all hypothesis testing discussed in the remainder of this chapter. For this reason we may present the results in an abbreviated form, using the results of chapter 6.

7.5 Testing the Variance and Standard Deviation of a Normal Distribution

To test the variance of a normal distribution using a sample of size n, refer to confidence interval results of section 6.4.2. For

$$H_0: \sigma^2 = \sigma_0^2, \qquad H_A: \sigma^2 \neq \sigma_0^2$$

the test statistic is

$$\chi^2 = \frac{(n-1)S^2}{\sigma_0^2}$$

which has a χ^2-distribution with $n - 1$ degrees of freedom. We reject H_0 at a significance level α if $\chi^2 > \chi^2_{\alpha/2}$ or if $\chi^2 < \chi^2_{1-\alpha/2}$.

If the alternative hypothesis is $H_A: \sigma^2 < \sigma_0^2$, H_0 is rejected for $\chi^2 < \chi^2_{1-\alpha}$. If the alternative hypothesis is $H_A: \sigma^2 > \sigma_0^2$, then H_0 is rejected for $\chi^2 > \chi^2_{\alpha}$.

Example 7.5 Suppose that the method of estimating farmers' crop losses has a variance σ^2 of $64\%^2$. A new method is claimed to have a smaller variance. To test this claim, we formulate the following null and alternative hypotheses:

$$H_0: \sigma^2 = 64, \qquad H_A: \sigma^2 < 64$$

and take a random sample of 30 farms from which we calculate $S^2 = 60$. We test the claim with a 0.1 level of significance. The value of the test statistic is

$$\chi^2 = \frac{(n-1)S^2}{\sigma_0^2} = \frac{(29)(60)}{64} = 27.19$$

and the value of $\chi^2_{.90}$ is 19.77. (Appendix Table 5). Since χ^2 is not less than $\chi^2_{.90}$, we cannot reject H_0. We have no evidence that the new method reduces the variance.

7.6 Testing the Difference of Two Means

In section 6.4.4 we considered problems where the experimenter wished to compare two means (the means of two populations). In this section we consider the problem of hypothesis testing of two means. If μ_1 and μ_2 are the means of two normal populations and we wish to test

$$H_0: \mu_1 = \mu_2, \qquad H_A: \mu_1 \neq \mu_2$$

using two independent samples of sizes n_1 and n_2, respectively, at significance level α (for which the variances σ_1^2 and σ_2^2 are assumed to be known), the test statistic is

$$z = \frac{\overline{X}_1 - \overline{X} - (\mu_1 - \mu_2)}{\sqrt{\dfrac{\sigma_1^2}{n_1} + \dfrac{\sigma_2^2}{n_2}}}$$

This can be simplified since, when the null hypothesis is true, $\mu_1 = \mu_2$ is the same as $\mu_1 - \mu_2 = 0$. The test statistic is

$$z = \frac{\overline{X}_1 - \overline{X}_2}{\sqrt{\dfrac{\sigma_1^2}{n_1} + \dfrac{\sigma_2^2}{n_2}}}$$

which has a standard normal distribution. We reject H_0 at a significance level α if $z < Z_{\alpha/2}$ or if $z > Z_{\alpha/2}$.

If the alternative hypothesis is $H_A: \mu_1 < \mu_2$, we reject H_0 if $z < -Z_\alpha$. If the alternative hypothesis is $H_A: \mu_1 > \mu_2$, we reject H_0 if $z > Z_\alpha$.

Example 7.6 Use example 6.10 where the weight at 20 weeks of two populations of broiler chickens is compared. Group 1 received a weight-increasing hormone and group 2 received the standard diet. The weights of the two groups are assumed to be normally distributed with variances $\sigma_1^2 = 0.8$, $\sigma_2^2 = 1.2$. We wish to test if the two groups have equal mean weights (in which case the weight-increasing hormone has no effect) versus the alternative that group 1 has a higher mean:

$$H_0: \mu_1 = \mu_2, \qquad H_A: \mu_1 > \mu_2$$

using a 0.05 level of significance. From group 1 a sample of 20 observations yields $\overline{X}_1 = 6.3$ and from group 2 an independent sample of 30 observations yields $\overline{X}_2 = 5.7$. The test statistic is

$$z = \frac{\overline{X}_1 - \overline{X}_2}{\sqrt{\dfrac{\sigma_1^2}{n_1} + \dfrac{\sigma_2^2}{n_2}}} = \frac{6.3 - 5.7}{\sqrt{\dfrac{0.8}{20} + \dfrac{1.2}{30}}} = 2.12$$

and $Z_\alpha = Z_{.05} = 1.65$. Since z is greater than 1.65, we reject H_0. We therefore conclude that the weight-increasing hormone is effective.

If the two variances are unknown but presumed to be equal, we may then proceed as follows. (As in section 6.4.4, we shall not consider what to do if the variances are unknown and unequal. The interested reader will find this discussed in Brownlee 1965.) To test $H_0: \mu_1 = \mu_2$ against $H_A: \mu_1 \neq \mu_2$, we obtain two independent samples and compute $\overline{X}_1$ and S_1^2 from the first and $\overline{X}_2$ and S_2^2 from the second. S_1^2 and S_2^2, the sample variances, are then combined to produce S^2, the pooled estimator of σ^2, the population variance, as follows:

$$S^2 = \frac{(n_1 - 1)S_1^2 + (n_2 - 1)S_2^2}{n_1 + n_2 - 2}$$

The test statistic is

$$t = \frac{\overline{X}_1 - \overline{X}_2}{S\sqrt{\dfrac{1}{n_1} + \dfrac{1}{n_2}}}$$

which has a t-distribution with $n_1 + n_2 - 2$ degrees of freedom. We reject H_0 at significance level α if $t < -t_{\alpha/2}$ or if $t > t_{\alpha/2}$.

If the alternative hypothesis is H_A: $\mu_1 > \mu_2$, we reject H_0 if $t > t_\alpha$. If the alternate hypothesis is H_A: $\mu_1 < \mu_2$, we reject H_0 if $t < t_\alpha$.

Example 7.7 Continuing with example 7.6, suppose that the variances of the two populations are unknown but presumed to be equal, and that the sample of 20 observations from population 1 yields $S_1^2 = 1.0$, while the sample of 30 observations from population 2 yields $S_2^2 = 1.5$. Test the previous hypothesis (H_0: $\mu_1 = \mu_2$, H_A: $\mu_1 > \mu_2$) with $\alpha = 0.05$.

The value of S^2 is 1.30. The test statistic is

$$t = \frac{\overline{X}_1 - \overline{X}_2}{S\sqrt{\frac{1}{n_1} + \frac{1}{n_2}}} = \frac{6.3 - 5.7}{(1.30)\sqrt{\frac{1}{20} + \frac{1}{30}}} = 1.60$$

and $t_{.05} \cong 1.96$. (Because the degrees of freedom are 48, which is considered large, we may use the normal distribution in place of the t-distribution.) Since t is not greater than $t_{.05}$, we cannot reject H_0. We therefore do not have evidence to say that the weight-increasing hormone is effective.

7.7 Testing the Difference of Two Means When the Observations Are Paired

As in section 6.4.4, some experiments yield paired data so that we have n pairs of observations. The n individuals or objects are presumed to be independent of each other, but the two observations within a pair are not independent. If the means of the two groups are μ_1 and μ_2 and the differences D_i for each pair are assumed to have a normal distribution, to test

$$H_0: \mu_1 = \mu_2, \qquad H_A: \mu_1 \neq \mu_2$$

the test statistic is

$$t = \frac{\overline{D}}{S_D/\sqrt{n}}$$

where

$$\overline{D} = \frac{\sum_{i=1}^{n} D_i}{n}, \text{ and } S_D^2 = \frac{\sum_{i=1}^{n} (D_i - \overline{D})^2}{n - 1}$$

The test statistic has a t-distribution with $n - 1$ degrees of freedom. We reject H_0 at significance level α if $t < -t_{\alpha/2}$ or $t > t_{\alpha/2}$.

If the alternative hypothesis is H_A: $\mu_1 < \mu_2$, we reject H_0 if $t < -t_\alpha$. If the alternative hypothesis is H_A: $\mu_1 > \mu_2$, we reject H_0 if $t > t_\alpha$.

Example 7.8 Consider the data of example 6.12 and use it to test H_0: $\mu_1 = \mu_2$, H_A: $\mu_1 \neq \mu_2$ when μ_1 and μ_2 are the mean responses of two laboratories. Use $\alpha = 0.05$.

Here $\overline{D} = 0.24$, $S_D = 0.83$. The test statistic is

$$t = \frac{\overline{D}}{S_D/\sqrt{n}} = \frac{0.24}{0.83/\sqrt{10}} = 0.914$$

and $t_{.025} = 2.26$. Since $t = 2.26$, we cannot reject H_0. The mean responses of the two laboratories are equal.

7.8 Testing a Proportion

If a discrete random variable has a binomial distribution, we may wish to test a hypothesis concerning the population proportion p. Consider

$$H_0\text{: } p = p_0 \qquad H_A\text{: } p \neq p_0$$

If from a random sample of size n we obtain X successes, the test statistic, from section 6.4.3, is

$$z = \frac{\frac{X}{n} - p_0}{\sqrt{\frac{p_0 (1 - p_0)}{n}}}$$

which has approximately a standard normal distribution (for large n). We reject H_0 at significance level α if $z < -Z_{\alpha/2}$ or if $z > -Z_{\alpha/2}$.

If the alternative hypothesis is H_A: $p < p_0$, we reject H_0 if $z < -Z_\alpha$. If the alternative hypothesis is H_A: $p > p_0$, we reject H_0 if $z > Z_\alpha$.

Example 7.9 Reconsider example 6.8 and suppose that it claims that no more than 10% of the consignment of apples is bad. We wish to test

$$H_0: p \leq 0.1, \qquad H_A: p > 0.1$$

at significance level 0.05 and select a random sample of n = 100 apples, of which X = 17 are bad. The test statistic is

$$z = \frac{0.17 - 0.10}{\sqrt{\dfrac{(0.1)(0.9)}{100}}} = 2.33$$

and $Z_{.05} = 1.65$. Since $z > Z_{.05}$, we reject the null hypothesis and conclude that the proportion of bad apples in the consignment is significantly greater than 10%.

7.9 Testing for Normality

Since the assumption regarding the normality of data is used very frequently, it is of interest to the researcher to have some techniques available for checking the validity and truth of this assumption. Though many methods are available to test normality of data, we shall restrict ourselves to the discussion of skewness and kurtosis. A more general test, known as the χ^2-*goodness of fit* test, is described in chapter 8 (example 8.5).

We know that the normal distribution is symmetric about its mean and is bell-shaped. Distributions that are not symmetrical are said to be *skewed.* Distributions that are too flat or too peaked are said to exhibit *kurtosis.* Nonnormality of a distribution can be investigated by studying the specific aspects of skewness and kurtosis.

7.9.1 Skewness

If a distribution is asymmetric, it is skewed. Skewness can be positive or negative. Figure 7.2 illustrates these two types of skewness.

A positively skewed distribution has a longer tail to the right and a negatively skewed distribution has a longer tail to the left. A symmetric distribution is said to have no skewness. To estimate and test skewness we need a random sample $X_1, X_2, \ldots, X_n$ from the population being inspected. If we denote the skewness of a population by γ_1 (gamma), an estimate of γ_1 is given by

$$g' = \frac{m_3}{m_2^{3/2}}$$

where $$m_2 = \frac{\Sigma (X_i - \overline{X})^2}{n}, \quad m_3 = \frac{\Sigma (X_i - \overline{X})^3}{n}$$

The distribution of g′ has been studied and the two critical values $g'_{.05}$ and $g'_{.01}$ are given in Appendix Table 7. To use the table validly, n must be at least 25. We can determine that a population is symmetric (i.e., one for which $\gamma_1 = 0$) by testing H_0: $\gamma_1 = 0$ against H_A: $\gamma_1 > 0$ at significance level 0.05 or 0.01 and comparing g′ to the appropriate tabulated value. If g′ is larger than the tabulated value, we reject H_0. To test H_0: $\gamma_1 = 0$, H_A: $\gamma_1 < 0$, we affix minus signs to the tabulated values in Appendix Table 7. If g′ is less than these negative quantities, we reject H_0. To use the two-sided test H_0: $\gamma_1 = 0$, H_A: $\gamma_1 \neq 0$, we call the tabulated value g'_α. If $g' > g'_\alpha$ or $g' < -g'_\alpha$, we reject H_0. However, the level of signifi-

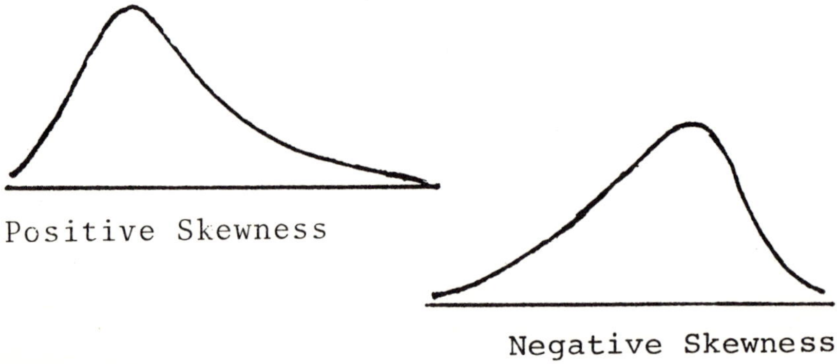

Fig. 7.2 *Skewed distributions.*

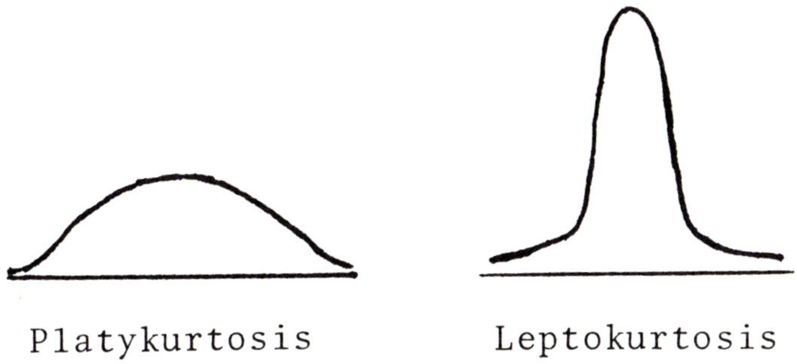

Fig. 7.3 *Types of kurtosis.*

cance will be 2α. Thus, a two-sided test can be run at significance level 0.02 (using the 0.01 value) and 0.10 (using the 0.05 value).

7.9.2. Kurtosis

Kurtosis describes the relative peakness or flatness of the distribution. There are two types of kurtosis, namely, platykurtosis and leptokurtosis, as illustrated in figure 7.3.

The normal distribution is referred to as mesokurtic. The kurtosis of a distribution is measured by a quantity called γ_2 which is estimated by

$$g'' = \frac{m_4}{m_2^2} - 3$$

where $$m_4 = \frac{\Sigma (X_i - \overline{X})^4}{n} \quad \text{and} \quad m_2 = \frac{\Sigma (X_i - \overline{X})^2}{n}$$

The value of γ_2 for a normal distribution is 0. If $\gamma_2 > 0$ the distribution is leptokurtic and if $\gamma_2 < 0$ it is platykurtic. Unfortunately, no tabular values are available for the distribution of g″. For this reason we give a better test of kurtosis devised by R. C. Geary, known as the g-test (Remington and Schork 1970). According to this test, we calculate the quantity g as follows:

$$g = \frac{\sum_{i=1} |X_i - \overline{X}|}{S\sqrt{n(n-1)}},$$

where S is the standard deviation of the data. The critical values for $g_{.99}$, $g_{.95}$, $g_{.90}$, $g_{.10}$, $g_{.05}$, $g_{.01}$ for the distribution of g for random samples from a normally distributed population are given in Appendix Table 8. These tabulated values allow us to carry upper or lower one-sided tests of kurtosis at the 1%, 5%, and 10% levels of significance, and two-sided tests at the 2%, 10%, and 20% levels of significance. Denoting Geary's population measure of kurtosis by γ_2, to test H_0: $\gamma_2 = 0$, H_A: $\gamma_2 > 0$ at an α level of significance, we reject H_0 if $g > g_\alpha$. To test H_0: $\gamma_2 = 0$, H_A: $\gamma_2 < 0$ at an α level of significance, we reject H_0 if $g < g_{1-\alpha}$. To test H_0: $\gamma_2 = 0$, H_A: $\gamma_2 \neq 0$ at a 2α level of significance, we reject H_0 if $g < g_{1-\alpha}$ or if $g > g_\alpha$.

Example 7.10 Consider the frequency distribution of table 2.2 regarding the percentage of crop loss data for 60 farmers given in table 2.1. Test the distribution of crop losses for normality by testing for skewness and kurtosis. In particular, test H_0: $\gamma_1 = 0$, H_A: $\gamma_1 > 0$ and then H_0: $\gamma_2 = 0$, H_A: $\gamma_2 > 0$, both tests at a 0.05 level of significance. Table 2.2 is reproduced here for convenience.

Frequency distribution for percentage of crop loss

X_i	f_i
5.5	2
7.5	5
9.5	10
11.5	20
13.5	14
15.5	6
17.5	3

For skewness, $\overline{X} = 11.8$ (from example 3.2)

$$m_2 = \frac{\Sigma (X_i - \overline{X})^2 f_i}{n}$$

$$= \frac{(5.5 - 11.8)^2 (2) + \ldots + (17.5 - 11.8)^2 (3)}{60} = 7.443$$

$$m_3 = \frac{\Sigma (X_i - \overline{X})^3 f_i}{n}$$

$$= \frac{(5.5 - 11.8)^3 (2) + \ldots + (17.5 - 11.8)^3 (3)}{60} = 32.469$$

so that $$g' = \frac{m_3}{m_2^{3/2}} = \frac{32.469}{(7.443)^{3/2}} = 1.598$$

and $g'_{.05}$, from Appendix Table 7 with n = 60 is 0.492. Since g′ is larger than the tabulated value, we reject H_0 and conclude that the distribution of crop losses is not symmetric but skewed. For kurtosis,

$$s = \sqrt{\frac{\Sigma (X_i - \overline{X})^2 f_i}{n - 1}} = 2.75 \text{ (from table 3.1)}$$

so that

$$g = \frac{\Sigma |X_i - \overline{X}| f_i}{s\sqrt{n(n - 1)}}$$

$$= \frac{|5.5 - 11.8| (2) + \ldots + |17.5 - 11.8| (3)}{(2.75) \sqrt{60 \times 59}} = 0.771$$

The value of $g_{.05}$ from Appendix Table 8 with n = 60 is approximately 0.7554. Since g is larger than $g_{.05}$, we reject H_0 and conclude that the distribution of crop losses is leptokurtic.

Exercises

1. An experiment is run to test the null hypothesis that a coin is fair (the probability of a head is 0.5). It is decided to test the hypothesis by flipping the coin 10 times and accepting H_0 if the number of heads is between 4 and 6 (inclusive). Otherwise the null hypothesis is rejected. Calculate the probabilities of type I and type II errors.

2. Weights of sacks of potatoes are assumed to be normally distributed with variance σ^2 equal to 25. We wish to test

H_0: $\mu \le 20$ against H_A: $\mu > 20$, where μ is the population mean weight with a 0.1 level of significance. To do this a sample of 20 observations is taken and the sample mean is calculated to be 21.0. (i) What decision is made? (ii) Calculate the power of the test if in fact $\mu = 20.5$ for samples of size 10, 20, 30, and 40. What happens to the power of the test as n increases?

3. Weights of cans of peaches are normally distributed with $\sigma^2 = 0.25$. We wish to test H_0: $\mu \neq 1$ at 0.05 level of significance. We take a sample of size n and calculate $\overline{X} = 1.1$. For n = 5, 19, 20, 30 calculate (i) the critical region and use the test statistic to make a decision, and (ii) the power of the test if in fact $\mu = 1.005$.

4. Volumes of bottles of ketchup are assumed to be normally distributed with unknown mean μ and unknown variance σ^2. To see if the volumes meet production standards, the quality control manager wishes to test that H_0: $\mu = 200$ cc against H_A: $\mu \neq 200$ cc at a significance level 0.10. To do this the manager takes a sample of 25 bottles and finds that $\overline{X} = 210$ and $S^2 = 30$. Test the hypothesis.

5. In exercise 4 test that H_0: $\sigma^2 \le 28$, H_A: $\sigma^2 > 28$ at the 5% level of significance.

6. Shelf life of a particular dairy product is assumed to be normally distributed with mean μ and variance σ^2 (both unknown). To test that H_0: $\sigma^2 \le 3$ against H_A: $\sigma^2 > 3$ at significance level 0.05, a random sample of 16 is taken and yields $S^2 = 4$. What decision is made?

7. In exercise 6 test that H_0: $\mu \le 10$ against H_A: $\mu > 10$ with $\alpha = 0.10$.

8. A fruit canning manufacturer claims that the sugar content of a particular product is less than 15%. Buyers, doubting this, decide to test H_0: $\mu \le 15$ against H_A: $\mu > 15$, where μ is the mean sugar content for the product population, using a level of significance of 0.10. They take a random sample of 50 cans and find that $\overline{X}$ and S^2 are 16 and 1, respectively. What decision do they come to?

9. In judging boxes of incoming fruit, we wish to see if the percentage of ripe fruit is too high (above 20%). To this end we decide to test H_0: $p \leq 0.2$ against H_A: $p > 0.2$ at significance level 0.10, where p is the population proportion of ripe fruit. We take a random sample of 100 of the fruit and find that 25 are ripe. Now test the hypothesis.

10. In a study of smoking habits, it is claimed that the proportion p of smokers in the population is less than 30%. It is decided to test that H_0: $p \leq 30\%$ against H_A: $p > 30\%$ by taking a sample of 1000 individuals. The test is to be run at the 0.1 level of significance. If 400 of the 1000 individuals smoke, what conclusion would you come to?

11. To compare the effects of two treatments on the yield of corn, 25 plots are given treatment 1 and 20 other plots are given treatment 2. The responses are assumed to have normal distributions with a priori known variances of $\sigma_1^2 = 5.3$ and $\sigma_2^2 = 8.4$, respectively. We wish to test H_0: $\mu_1 = \mu_2$ against H_A: $\mu_1 > \mu_2$ at the 0.1 level of significance. If the sample means for the two treatments are 83 and 64, respectively, what would you conclude?

12. In exercise 11 suppose that the variances are unknown but assumed to be equal and are estimated from the samples to be $S_1^2 = 7.3$ and $S_2^2 = 9.8$. What conclusion would be reached?

13. A canning plant's production workers are given a training program and, to see if it is effective, their time for performing a particular task is measured in seconds before and after the program. We wish to test the effectiveness of the program by testing H_0: $\mu_B - \mu_A$ against H_A: $\mu_B > \mu_A$ (where μ_A and μ_B are the mean times for the task after and before training) at significance level 0.05. The differences (before − after) are as follows: 19, 2, 17, −4, 16, 24, 13, −10, −7, 4, 12, −6, 15. What decision is reached?

8.

Applications of the Chi-Square (χ^2) Distribution

Observational data from an experiment frequently take the form of frequencies or counts of separate individuals or units classified into utually exclusive categories such as number of diseased apple trees in an orchard or percentage of seeds germinated. The statistic called chi-square (χ^2) is generally used to analyze qualitative or enumerative data of this type. The test provides a means of determining whether the divergence of facts from a hypothesis is significant or not. The chi-square test is applied in two different ways:

1. Tests of goodness of fit. In problems of this type, the observed frequencies in a distribution are compared with those expected under the hypothesis to test whether their discrepancies are greater than could reasonably be attributed to chance fluctuations in the observed data. The theoretical distribution may be binomial, Poisson, normal, or hypothetical.

2. Tests of independence or homogeneity. If individuals in a sample are classified in two or more ways, we can test the hypothesis that these classifications are independent, i.e., there is no association between these attributes. Alternatively, we may test the hypothesis that the numbers in the two classifications are not significantly different and can be regarded as samples drawn from the same population.

8.1 The Chi-Square (χ^2) Test

For the solution of these and similar problems we need: (i) a measure of divergence of fact from hypothesis, and (ii) a means of judging whether or not the divergence is significant. The hypothesis to be tested is that the k observed frequencies $O_1, O_2, \ldots, O_k$ conform to a model that would produce expected frequencies $E_1, E_2, \ldots, E_k$. If the difference between each observed frequency and the corresponding expected frequency is squared and divided by the expected frequency, the sum of these terms is the value of χ^2 as follows:

$$\chi^2 = \sum_{i=1}^{k} \frac{(O_i - E_i)^2}{E_i}$$

The value of χ^2 will be 0 if the observed and expected frequencies completely coincide. As the value of χ^2 increases, the observed frequencies depart more and more from the expected ones. The sampling distribution of this χ^2 statistic is closely approximated by the chi-square distribution discussed in chapter 5. To determine whether or not the divergence between observed and expected frequencies is due to chance, the calculated value of χ^2 is compared with the χ^2-values (Appendix Table 5) for the given degrees of freedom at a certain level of significance. If the calculated value of χ^2 is less than the table value, the divergence between the observed and the expected value is said to arise from sampling fluctuations. If the calculated value of χ^2 exceeds the table value, the divergence between the observed and the expected frequencies is said to be significant.

The number of degrees of freedom is given by $\upsilon = k - 1 - m$ where m is the number of parameters in the model which must be estimated from the data. The maximum number of degrees of freedom is $k - 1$, obtained when $m = 0$.

Before using the χ^2 distribution as a test of significance, we must heed the following precautions:

1. The total frequency N should be fairly large. It should be at least 50, however small the number of categories may be.
2. Each observation must be independent of all other observations.

3. No expected category frequency should be small. It should not be less than 5 and it is better if the smallest frequency is 10 or greater. When the category frequencies are small in two or more cells, they should be combined into a single cell.

4. The number of classes of categories, k, should be neither too small nor too large. It is better if $5 \leq k \leq 20$. Sometimes fewer than 5 categories can also be used, provided the frequencies are not small.

Example 8.1 In pea breeding experiments, Mendel obtained the following frequencies of seeds:

Round and yellow	Wrinkled and yellow	Round and green	Wrinkled and green	Total
315	101	108	32	556

Theory predicts that the frequencies should be in proportions 9:3:3:1. Examine the correspondence between theory and experiment using a 5% level of significance.

The corresponding expected frequencies are

$$\frac{9}{16} \times 556, \frac{3}{16} \times 556, \frac{3}{16} \times 556, \frac{1}{16} \times 556$$

or 313, 104, 104, and 35, respectively.
Hence:

$$\chi^2 = \frac{(315 - 313)^2}{313} + \frac{(101 - 104)^2}{104} + \frac{(108 - 104)^2}{104} + \frac{(32 - 35)^2}{35}$$

$$= 0.51$$

$$\upsilon = 4 - 1 = 3$$

The value of χ^2 for 3 degrees of freedom at a 5% level of significance is $\chi^2_{0.05} = 7.81$ (from Appendix Table 5). Since the calculated value of $\chi^2 = 0.5$ is much less than the table value of $\chi^2_{0.05} = 7.81$, there is very high degree of agreement between theory and experiment.

Example 8.2 The number of reversals of potato shipments in a one-year period for the four Canadian potato-growing provinces is as follows:

Province	Number of reversals
1	11
2	12
3	9
4	16
Total	48

Test the hypothesis that each province has the same rate of reversals using a 5% level of significance.

On the hypothesis that each province has the same rate of reversals, the expected number of reversals in each province is 12. The χ^2 statistic is calculated as in table 8.1.

Table 8.1 *Expected reversals by province.*

Province	Observed frequency O_i	Expected frequency E_i	$O_i - E_i$	$\frac{(O_i - E_i)^2}{E_i}$
1	11	12	−1	0.08
2	12	12	0	0.00
3	9	12	−3	0.75
4	16	12	+4	1.33
Total	48	48		$\chi^2 = 2.16$

The number of degrees of freedom is

$$\upsilon = k - 1 = 4 - 1 = 3$$

The value of $\chi^2_{0.05}$ for 3 degrees of freedom = 7.81. Since the calculated value of $\chi^2 = 2.16$ is less than the table value of $\chi^2_{0.05} = 7.81$, the hypothesis that the four provinces have an equal rate of reversals cannot be rejected.

8.2 χ^2 as a Test of Goodness of Fit

One of the principal uses of the chi-square distribution is to test how well an observed distribution fits a theoretical distribution. Used in this way, it is called a test of "goodness of fit," and may be used in two ways, either to test the fit of the observed data to the hypothetical data or to test the merits of a particular formula or a particular curve in graduating a set of values without reference to a hypothesis. For example, we may test how well a normal curve or a Poisson curve fits the given data. Calculations in both cases are exactly on the same lines. We have already given a few examples for testing the goodness of fit of observed to hypothetical data. In these cases the population was completely specified, there being no unknown parameters in its distribution. The only limitation was that the tables of the observed and hypothetical data had to agree. We now add a few more examples for testing the goodness of fit where the expected frequencies are not completely determined by the hypothesis, but depend on some unknown parameters. This is the case when the hypothesis states that the population is normal without specifying μ and σ. As before, we may calculate the probability P to find how good the fit is. We generally regard low values of P as describing a poor fit. Very high values of P indicate an excellent fit, provided we apply the term to the closeness of the curve to the data. But when we are testing a certain hypothesis, a very high value of P does not necessarily establish the truth of the hypothesis. In these cases the hypothesis is as strongly disproved as in the case when the value of P is very low.

Example 8.3 In a germination trial, 20 seeds were grown in each of 100 plates. After 24 hours the number of germinated seeds in each plate was counted. The distribution is as follows:

No. of seeds germinated	x:	0	1	2	3	4	5	6	7	8	9	Total
No. of plates	f_o:	3	8	11	15	16	14	12	11	9	1	100

Fit a binomial distribution to the data and test its goodness of fit using a 5% level of significance.

The total number of seeds put into plates was $20 \times 100 = 2000$. Of these, the number of seeds germinated was $\Sigma X\, f_o = 439$. On the hypothesis that each seed has an equal chance of germinating, the proportion of the germinated seeds would be $\frac{439}{2000} = 0.2195$. Hence, the binomial distribution fitted to the data is

$$100\binom{20}{x}(0.2195)^{x}\,(6.7805)^{20-x}$$

The successive terms of this expansion give the expected frequencies as calculated here:

X:	0	1	2	3	4	5	6	7	8	9	10	11	Total
O_i:	3	8	11	15	16	14	12	11	9	1	0	0	100
E_i:	0.7	4.0	10.6	17.8	21.3	19.2	13.5	7.6	3.5	1.3	0.4	0.1	100

We regroup the frequencies so that no class contains fewer than five. Regrouping is shown by brackets. The calculations for the χ^2 statistic are shown in table 8.2.

Table 8.2 χ^2 *Statistic for example 8.3.*

No. of seeds germinated X	Observed frequency O_i	Expected frequency E_i	$O_i - E_i$	$\frac{(O_i - E_i)^2}{E_i}$
0 to 2	22	15.3	6.7	2.93
3	15	17.8	−2.8	0.44
4	16	21.3	−5.3	1.32
5	14	19.2	−5.2	1.41
6	12	13.5	−1.5	0.17
7	11	7.6	3.4	1.52
8, 9, 10 and 11	10	5.3	4.7	4.17
Total	100	100		$\chi^2 = 11.96$

Since the parameter p, proportion germinated, was calculated from the sample, the number of degrees of freedom $\upsilon = 7 - 1 - 1 = 5$. The value of $\chi^2_{.05}$ for $\upsilon = 5$ (from Appendix Table 5) is found to be 11.07. Since the calculated value is greater than this value, we conclude that the binomial distribution does not give a good fit to the observed data.

Example 8.4 Consider the data of table 2.2 (example 2.1, chapter 2) for the percentage of crop loss due to drought, as reported by 60 farmers in an area. Using a 5% level of significance, examine the goodness of fit of the normal distribution to the data.

The expected frequencies are calculated using Appendix Table 3 which gives the areas under the normal probability curve. To provide the relevant areas for probabilities corresponding to each class interval, we must first convert the class boundaries to values of z, the standard normal variable, given by $z = \frac{X - \overline{X}}{S}$. X is a boundary, and $\overline{X}$ and S are the sample mean and standard deviation used as estimates of their corresponding population parameters μ and σ.

The values of $\overline{X}$ and S for this frequency distribution are given by

$$\overline{X} = 11.8 \text{ and } S = 2.75$$

The probability, under the normal curve, of an individual falling in the first class is equal to the area to the left of 6.5. Similarly, the probability that an individual will fall in the last class is equal to the area to the right of 16.5. The probability of an individual falling in any other class can be found by calculating the z-value for lower and upper class limits, and then finding the area between the two z-values. Calculations of these probabilities and their corresponding expected frequencies are shown in table 8.3.

Table 8.3 *Probability and frequency calculations for example 8.4.*

Class Limits	Observed frequency O_i		$Z = \frac{X - \overline{X}}{S}$	Probability	Expected frequency E_i	
4.5–6.5	2	7	−1.9273	0.0268	1.61	6.91
6.5–8.5	5		−1.2000	0.0883	5.30	
8.5–10.5	10		−0.4727	0.2041	12.25	
10.5–12.5	20		0.2545	0.2795	16.76	
12.5–14.5	14		0.9818	0.2378	14.27	
14.5–16.5	6	9	1.7191	0.1208	7.25	9.81
16.5–18.5	3		2.4364	0.0427	2.56	
Total	60			1.0000	60	

The χ^2 statistic is calculated in table 8.4.

Table 8.4 χ^2 *Calculations for example 8.4.*

Class	Observed frequency O	Expected frequency E	O − E	$\frac{(O-E)^2}{E}$
1 and 2	7	6.91	0.09	0.0011
3	10	12.25	−2.25	0.4132
4	20	16.76	3.24	0.6263
5	14	14.27	−0.27	0.0051
6 and 7	9	9.81	−0.81	0.0669
Total	60	60		χ^2 = 1.1126

Since the parameters μ and σ are estimated from the sample, the number of degrees of freedom is $\upsilon = 5 - 2 - 1 = 2$. The χ^2 value for 2 degrees of freedom at a 5% level of significance (from Appendix Table 5) is 5.99. Since the calculated value of $\chi^2 =$ 1.1126 is less than the table value, we accept the hypothesis that the data follow a normal distribution. Thus, the fit of the normal distribution to the data is good.

Example 8.5 The following table gives the number of weed seeds in 196 1-lb packets of a variety of beans. Fit a Poisson distribution and test the goodness of fit.

Number of weed seeds	Observed frequency
0	7
1	33
2	54
3	37
4	34
5	16
6	8
7	5
8	1
9	1
10 or more	0
Total	196

The expected frequencies are calculated using the Poisson distribution by:

$$f(x) = \frac{e^{\lambda} (\lambda)^{x}}{x!}, x = 0, 1, 2 \ldots$$

The estimate of the mean (λ) is given by $\overline{X}$ calculated from the sample as $\overline{X} = 2.9$. The expected frequencies are calculated by N f(x), x = 0, 1, 2, . . . where N is the total sample size, in this case 196. The expected frequencies and the value of the χ^2 statistic are calculated in table 8.5.

Table 8.5 *Frequencies and χ^2 statistic for example 8.5.*

Number of weed seeds	Observed frequency O		Expected frequency E		O − E	$\frac{(O - E)^2}{E}$
0	7		10.78		−3.78	1.3255
1	33		31.28		1.72	0.0946
2	54		45.35		8.65	1.6499
3	37		43.84		−6.84	1.0672
4	34		31.78		2.22	0.1551
5	16		18.43		−2.43	0.3204
6	8		8.91		−0.91	0.0929
7	5		3.69			
8	1	=7	1.34	=5.63	1.37	0.3334
9	1		0.43			
10 or more	0		0.17			
Total	196		196		χ^2	= 5.039

Since the parameter λ is estimated from the sample, the number of degrees of freedom is $\upsilon = 8 - 1 - 1 = 6$. From Appendix Table 5, the value of χ^2 for 6 degrees of freedom at a 5% level of significance is 12.5916. Since the calculated value of $\chi^2 = 5.039$ is less than this value, the hypothesis that the frequency distribution of weed seeds follows a Poisson distribution is accepted and the Poisson fit is good.

8.3 Test of Independence: Contingency Tables

A special and important class of cases where compatibility of observed and expected frequencies has to be tested occurs when the independence of two attributes is tested from data arranged in two-way tables called *contingency tables*. This classification of individuals may be multiple or dichotomous. The hypothesis to be tested is that there is an association between the row and column classification variables or that the two classifications of variables are independent. For the hypothesis of independence we calculate the value of χ^2. If the χ^2-test shows that the divergence between observation and expectation is significant, we have to reject the hypothesis. For example, an animal may be classified as inoculated or not inoculated, and also as attacked and not attacked by a disease. We may need to know if the two classifications are independent. A very useful application of the χ^2-test is to investigate the relationship between traits or attributes that can be classified into two or more categories.

To develop the contingency table test procedure, we consider a rectangular arrangement with r rows and c columns, called an r × c contingency table. The rows represent r classes $X_1, X_2, \ldots, X_r$ of attribute X, and c classes $Y_1, Y_2, \ldots, Y_c$ of attribute Y. The observed frequency in row i and column j, the (i,j) cell of the table, is denoted by O_{ij}. The row marginal totals are denoted by $N_{1.}, N_{2.}, \ldots, N_{r.}$, the column marginals by $N_{.1}, N_{.2}, \ldots, N_{.c}$. N denotes the total sample size.

Table 8.6 *r × c Contingency table.*

	Columns			
Rows	Y_1	Y_2	Y_c	Row totals
X_1	O_{11}	$O_{12} \ldots$	O_{1c}	$N_{1.}$
X_2	O_{21}	$O_{22} \ldots$	O_{2c}	$N_{2.}$
.		.	.	.
.	.	.	.	.
.	.	.	.	.
X_r	O_{r1}	O_{r2}	O_{rc}	$N_{r.}$
Column totals	$N_{.1}$	$N_{.2} \ldots$	$N_{.c}$	N

To test whether there is any relationship between attributes X and Y, we set up the null hypothesis of no relationship, or independence of X and Y, and calculate the expected frequencies in all the categories on this hypothesis. Since we wish to test independence only and not any hypothesis as to the total numbers belonging to different classes, the expected frequencies are calculated by keeping the class totals fixed. This is equivalent to imposing linear constraints that the class totals of the observed and expected frequencies must agree.

Let E_{ij} denote the expected frequency in the cell common to X_i and Y_j. Then, since X and Y are independent, in any class the proportions of members belonging to $X_1, X_2, \ldots, X_r$ should be the same and equal to the proportions in the total. Thus,

$$\frac{E_{ij}}{N_{1.}} = \frac{E_{2j}}{N_{2.}} = \ldots = \frac{E_{rj}}{N_{r.}} = \frac{N_{.j}}{N}$$

so that $E_{ij} = \dfrac{N_{i.} \times N_{.j}}{N}$

On the basis of independence the expected frequency in any category is equal to the product of the class totals of the two classes to which the cell belongs, divided by the total number of observations. Hence,

$$\chi^2 = \sum_{i=1}^{r} \sum_{j=1}^{c} \frac{(O_{ij} - E_{ij})^2}{E_{ij}}$$

$$= \sum_{i=1}^{r} \sum_{j=1}^{c} \frac{\left[O_{ij} - \dfrac{N_{i.} \times N_{.j}}{N}\right]^2}{\dfrac{N_{i.} \times N_{.j}}{N}}$$

The number of degrees of freedom for an r × c contingency table is equal to the number of cells whose frequencies can be arbitrarily assigned. Since the expected frequencies in all cells except in the last row and the last column can be determined given the row totals, column totals, and the other cell frequencies, the number of degrees of freedom is

$$\upsilon = rc - (r - 1) - (c - 1) = (r - 1)(c - 1)$$

Example 8.6 Table 8.7 gives data on the number of poppy plants destroyed and not destroyed by three different treatments given to 3 experimental plots of oats each containing 50, 80, and 100 poppy plants.

Table 8.7 *Response to three treatments.*

	Treatment			
Response	I	II	III	Total
Destroyed	35	50	60	145
Not destroyed	15	30	40	85
Total	50	80	100	230

Test the data to see whether or not there are differences in the responses to the different treatments.

On the hypothesis that there is no difference among the treatments, the expected frequencies may be calculated as:

$$E_{11} = \frac{145 \times 50}{230} = 31.52, \; E_{12} = \frac{145 \times 80}{230} = 50.43,$$

$$E_{13} = \frac{145 \times 100}{230} = 63.04, \; E_{21} = \frac{85 \times 50}{230} = 18.48,$$

$$E_{22} = \frac{85 \times 80}{230} = 29.56, \; E_{23} = \frac{85 \times 100}{230} = 36.96$$

The χ^2 statistic is calculated in Table 8.8.

The number of degrees of freedom for a 2 × 3 contingency table is

$$\upsilon = (r - 1)(c - 1) = (2 - 1)(3 - 1) = 2$$

The χ^2 value for 2 degrees of freedom at a 0.05 level of significance (from Appendix Table 5) is $\chi^2 = 5.99$. The calculated value

Table 8.8 χ^2 *Statistic for example 8.6.*

Classification	Observed frequency O_i	Expected frequency E_i	$O_i - E_i$	$\frac{(O_i - E_i)^2}{E_i}$
Destroyed and treatment I	35	31.52	3.48	0.3842
Destroyed and treatment II	50	50.43	−0.43	0.0037
Destroyed and treatment III	60	63.05	−3.05	0.1475
Not destroyed and treatment I	15	18.48	−3.48	0.6553
Not destroyed and treatment II	30	29.56	0.44	0.0065
Not destroyed and treatment III	40	36.96	3.04	0.2500
Totals	230	230		$\chi^2 = 1.4472$

of $\chi^2 = 1.4472$ is less than the table value of $\chi^2 = 5.99$. We have no reason to reject our hypothesis of no difference among treatments.

Example 8.7 In an experiment with immunization of cattle from tuberculosis, the following results were obtained:

	Affected	Not affected	Total
Inoculated	12	26	38
Not inoculated	16	6	22
Total	28	32	60

Examine the effect of vaccine in controlling susceptibility to tuberculosis.

Under the hypothesis that vaccine has no effect in controlling susceptibility to tuberculosis, the expected frequencies may be calculated as follows:

$$E_{11} = \frac{38 \times 28}{60} = 18, \qquad E_{12} = \frac{38 \times 32}{60} = 20$$

$$E_{21} = \frac{22 \times 28}{60} = 10, \qquad E_{22} = \frac{22 \times 32}{60} = 12$$

Hence,

$$\chi^2 = \sum \frac{(O - E)^2}{E}$$

$$= \frac{(12 - 18)^2}{18} + \frac{(26 - 20)^2}{20} + \frac{(16 - 10)^2}{10} + \frac{(6 - 12)^2}{12}$$

$$= 7.1$$

The number of degrees of freedom for a 2 × 2 contingency table is

$$\upsilon = (r - 1)(c - 1) = (2 - 1)(2 - 1) = 1$$

Using a 0.05 level of significance, the value of χ^2 for 1 degree of freedom (from Appendix Table 5) is 3.84. Since the calculated value of $\chi^2 = 7.1$ is greater than the table $\chi^2 = 3.84$, we reject the hypothesis that vaccine has no effect in controlling susceptibility to tuberculosis.

8.3.1 Further Consideration of 2 × 2 Contingency Tables

Most often the enumeration data lend themselves to the form of a 2 × 2 contingency table. Calculations of the χ^2 statistic for this type of table can be simplified as follows.

Consider the general form of the 2 × 2 contingency table 8.9:

Table 8.9 *The general 2 × 2 contingency table.*

	A	Not A	Total
B	a	b	a + b
Not B	c	d	c + d
Total	a + c	b + d	N

where a sample of size N = (a + b + c + d) has been classified in two ways according to (i) whether they do or do not possess the attribute A, and (ii) whether they do or do not possess the attribute B.

Since the marginal totals are fixed, the probability of a member

belonging to class A is $\frac{a + c}{N}$ and of it belonging to class B is $\frac{a + b}{N}$.
Further, the attributes being independent, the probability of it belonging to both A and B is $\left(\frac{a + b}{N}\right) \times \left(\frac{a + c}{N}\right)$ and, therefore, the expected frequency of this class, denoted by E(a), is given by

$$E(a) = N \times \frac{a + b}{N} \times \frac{a + c}{N} = \frac{(a + b)(a + c)}{N}$$

Similarly, $$E(b) = \frac{(a + b)(b + d)}{N}, \qquad E(c) = \frac{(c + d)(a + c)}{N},$$

$$E(d) = \frac{(c + d)(b + d)}{N}$$

and
By definition, $$\chi^2 = \sum \frac{(O_i - E_i)^2}{E_i}$$

$$= \frac{[a - E(a)]^2}{E(a)} + \frac{[b - E(b)]^2}{E(b)} + \frac{[c - E(c)]^2}{E(c)} + \frac{[d - E(d)]^2}{E(d)}$$

By simplifying, it can be shown that χ^2 is given by the formula

$$\chi^2 = \frac{N(ad - bc)^2}{(a + b)(a + c)(b + d)(c + d)}$$

The number of degrees of freedom for a 2×2 contingency table is

$$\upsilon = (r - 1)(c - 1) = (2 - 1)(2 - 1) = 1$$

8.4 Yates' Correction for Continuity

We stated in Section 8.1 that, for the χ^2 approximation to be valid, the expected frequency in each class must be sufficiently large. When some expected frequency is too small, we coalesce some of the classes in order to satisfy this condition. However, this procedure is ruled out in the case of testing for independence in a 2 × 2 table.

Yates suggested that a correction be applied to the observed frequencies in a 2 × 2 table when any expected frequency is too small. This correction consists in increasing or decreasing the observed frequencies by ½ in such a way that the marginal totals remain unaltered. If we consider the 2 × 2 table in section 8.3.1, we must increase a and d by ½ each (while each of b and c is decreased by ½) in case ad < bc. On the other hand, if ad > bc, a and d have to be decreased by ½ each, while each of b and c has to be increased by the same amount. Since the marginal totals remain unaltered, the expected frequencies are not changed.

If Yates' correction is applied, the formula for χ^2 corresponding to the one given in section 8.3.1 is

$$\chi^2 = \frac{N\left(|ad - bc| - \frac{N}{2}\right)^2}{(a + b)(a + c)(b + d)(c + d)}$$

with 1 degree of freedom.

Example 8.8 From the following table showing the number of plants having certain characters, test the hypothesis that flower color is independent of leaf flatness.

	Flat leaves	Curled leaves	Total
White flowers	45	32	77
Red flowers	20	7	27
Total	65	39	104

On the hypothesis that flower color is independent of leaf flatness, the value of χ^2 can be calculated using the formula of section 8.3.1.

Thus,

$$\chi^2 = \frac{N(ad - bc)^2}{(a + b)(a + c)(b + d)(c + d)}$$

$$= \frac{104[(45 \times 7) - (32 \times 20)]^2}{77 \times 65 \times 39 \times 27} = 2.08$$

The number of degrees of freedom is $\upsilon = (2 - 1)(2 - 1) = 1$. The 5% value of χ^2 for $\upsilon = 1$ is 3.841. Since the calculated value of χ^2 is less than this value, the result is not significant and the hypothesis of independence between leaf flatness and flower color cannot be rejected.

Example 8.9 Inspection of 200 cans of pears produced in a food plant gave the following results. Test the hypothesis that the fill weight of a can is independent of its being mutilated.

	Mutilated	Not mutilated	Total
Underfilled	50	40	90
Overfilled	20	90	110
Total	70	130	200

Under the hypothesis that mutilation of cans does not affect the fill weight of cans, the value of χ^2 can be calculated as follows:

$$\chi^2 = \frac{200[(50 \times 90) - (40 \times 20)]^2}{90 \times 70 \times 130 \times 110} = 30.4$$

The number of degrees of freedom for a 2×2 table is $\upsilon = (2 - 1)(2 - 1) = 1$. Using a 5% significance level, the value of χ^2 for 1 degree of freedom (from Appendix Table 5) is 3.841. Since the calculated value of $\chi^2 = 30.4$ is greater than the table value of 3.841, the result is very significant. The hypothesis that the mutilation of cans does not affect the fill weight of cans is rejected.

Example 8.10 In a poultry experiment two new drugs for producing eggs containing yolks only were tested. The drugs were claimed to induce hens to lay eggs without albumen within 30 hours of administration. The results are as follows:

	No. of eggs with albumen and yolk	No. of eggs with yolk only	Total
Drug A	2	13	15
Drug B	5	14	19
Total	7	27	34

Test the efficiency of the two drugs.

Under the hypothesis that the two drugs are equally effective, the value of χ^2 is calculated as follows:

$$\chi^2 = \frac{34[(2 \times 14) - (13 \times 5)]^2}{7 \times 15 \times 27 \times 19} = 0.86$$

The number of degrees of freedom is $\upsilon = (2 - 1)(2 - 1) = 1$. The χ^2-value for 1 degree of freedom at the 5% significance level is 3.841 (from Appendix Table 5). Since the calculated value of $\chi^2 = 0.86$ is less than this value, we accept the hypothesis. The two drugs are equally effective in their ability to induce yolk-producing properties.

In this example, since the total sample size is less than 40 and two of the cell frequencies have expected frequencies less than 5, we should apply the χ^2 formula with Yates' continuity correction as described in section 8.4. Thus,

$$\chi^2 = \frac{N\left(|ad - bc| - \frac{N}{2}\right)^2}{(a + b)(a + c)(b + d)(c + d)}$$

$$= \frac{34\left(|2 \times 14 - 5 \times 13| - \frac{34}{2}\right)^2}{15 \times 7 \times 27 \times 19} = 0.25$$

As before, since the calculated value of $\chi^2 = 0.25$ is less than the tabulated value of $\chi^2_{0.05} = 3.841$ for 1 degree of freedom, we accept the hypothesis of no difference between the effectiveness of the two drugs. It may be noticed, however, that the continuity correction improves the results.

8.5 Fisher's Exact Test for 2 × 2 Tables

Testing independence by χ^2 tests in a 2 × 2 table is a useful approximation for the comparative simplicity of the calculations. The exact treatment is somewhat laborious, though necessary in case of doubt and valuable for displaying the true nature of the inference that the χ^2 method is designed to draw. Fisher's exact test is specially designed for a 2 × 2 table. It should be applied in all cases where the total frequency is less than 40 or any expected cell frequency is less than 5. This method computes the probability of obtaining the observed results and all worse cases. These probabilities are computed assuming that the row and column classifications are independent (the null hypothesis) and that the row and column totals are fixed.

Consider the general 2 × 2 contingency table in section 8.3.1 (Table 8.9).

	A	Not A	Total
B	a	b	a + b
Not B	c	d	c + d
Total	a + c	b + d	N = a + b + c + d

Fisher's exact test answers the following question: Given two-way tables with the same fixed marginal totals as the observed one, what is the chance of obtaining the observed cell frequencies that represent a greater deviation from expectation?

The number of different ways in which the body of this table can be completed is

$$\frac{N!}{a!b!c!d!}$$

and, assuming independence of two attributes, the total number of ways in which marginal totals can be obtained is

$$\frac{N!}{(a + c)!(b + d)!} \times \frac{N!}{(a + b)!(c + d)!}$$

Consequently, the probability of the distribution of the table is

$$P = \frac{(a + b)!(c + d)!(a + c)!(b + d)!}{a!b!c!d!N!}$$

From here we can find the probabilities for a = 0, 1, 2, . . . i.e., the probability of all worse cases. This is done by listing systematically all the worse cases, computing the probability of each, and summing these probabilities. Since we are usually interested in a two-tailed test, we should consider all worse cases in both tails, but as a computational convenience, we usually compute the probability for the tail that contains the observed result and double this probability.

Example 8.11 Consider example 8.10 and analyze the data using Fisher's exact test.

	No. of eggs with albumen and yolk	No. of eggs with yolk only	Total
Drug A	2	13	15
Drug B	5	14	19
Total	7	27	34

Under the hypothesis that the two drugs are equally effective, the probability of the distribution of the table is given by:

$$p_2 = \frac{(a + b)!\,(c + d)!\,(a + c)!\,(b + d)!}{a!b!c!d!N!}$$

$$= \frac{15!19!7!27!}{2!13!5!14!34!} = 0.2270$$

The subscript to p is the smallest of the quantities a, b, c, and d. In the present example, the smallest entry is in the top left-hand cell of the table and is 2.

Returning to the table and keeping in mind that the marginal frequencies are to be taken as fixed, the frequencies in the body of the table can be arranged in two other ways, both of which would represent more extreme discrepancies between the two groups. The possible arrangements are

More extreme cell frequencies than those observed

(a)

		Total
1	14	15
6	13	19
7	27	34

(b)

		Total
0	15	15
7	12	19
7	27	34

As before, the probabilities for these tables are calculated and we get the following:

$$p_1 = 0.03895 \text{ for table (a),}$$

$$\text{and } p_0 = 0.0006265 \text{ for table (b).}$$

The probability P of obtaining the observed cell frequencies or frequencies more extreme than the observed ones is therefore the sum of the three probabilities:

$$\begin{aligned} P &= p_2 + p_1 + p_0 \\ &= 0.2270 + 0.03895 + 0.0006265 \\ &= 0.2665765 \end{aligned}$$

For a two-sided test, this probability is doubled. Using a 5% significance level, we observe that the calculated value of probability is significantly greater than 0.05, indicating a high probability of accepting the hypothesis. The two drugs do not differ in their effectiveness in producing eggs containing yolk only.

Note: Simplifying the calculations
The tedium of calculations when using Fisher's exact test can be reduced by the use of logarithms. Appendix Table 14 gives logarithmic values of factorials. Using this table the value of p_2 can be calculated as follows:

$$p_2 = \frac{15!19!7!27!}{2!13!5!14!34!}$$

$$\begin{aligned} \log p_2 &= (\log 15! + \log 19! + \log 7! + \log 27!) \\ &\quad - (\log 2! + \log 3! + \log 5! + \log 14! + \log 34!) \\ &= (12.1165 + 17.0851 + 3.7024 + 28.0370) \\ &\quad - (0.3010 + 9.7943 + 2.0792 + 10.9404 + 38.4702) \\ &= -0.6441 = \overline{1}.3559 \end{aligned}$$

$$p_2 = \text{antilogarithm of } \overline{1}.3559 = 0.2270$$

In the same way, the values of p_1 and p_0 can also be solved.

Exercises

1. Five different makes of an agricultural implement were tried at the same time over a period of three months. The numbers of recorded breakage for each type were 15, 12, 13, 10, 14. Are there real differences between them in respect to breakage?

2. Genetic theory states that children having one parent of blood type M and the other of blood type N will always be one of the three types; M, MN, or N, and the proportions of the three types will, on the average, be 1:2:1. In a study of 250 children having one M parent and one N parent, 32% were found to be type M, 44% type MN, and the remainder type N. Test the hypothesis by the χ^2-test.

3. Referring to exercise 3, chapter 2 giving data on percent defective apples found during inspection of 6 lots, test the hypothesis that each lot has the same rate of defective apples, using a 5% level of significance.

4. Examine the goodness of fit of the normal distribution to the corn yield data of exercise 3, chapter 3.

5. The following table shows the results of inoculation against polio:

	Not attacked	Attacked
Inoculated	325	21
Not inoculated	173	29

Is there any significance between inoculation and attack?

6. The following table shows the result of inspection of 75 boxes of pears each containing 1 dozen pears.

Number of diseased pears	0	1	2	3	4
Number of boxes (observed frequency)	42	19	8	4	2

Fit a binomial distribution to the data and test its goodness of fit.

7. A clinical trial was conducted to compare the effectiveness of a placebo and an active drug (called injections A and B,

respectively) in stopping bleeding during tonsillectomy. Eighty children of both sexes between the ages of 2 and 11 were treated alternatively with injections A and B. Reactions to the injections during surgery were divided into the following three categories:

Bleeding	Represented by
Stopped completely	0
Stopped a little bit	+
Did not stop	++

The data are presented in the following table:

		Categories		
		0	+	++
Injections	A	8	10	4
	B	35	16	7

Test for independence.

8. The following table gives the frequency distribution of eggs of the cabbage root fly (*Hylemyia brassicae*) around rutabagas. Test the goodness of fit of a Poisson distribution to the data.

Number of eggs:	0	1	2	3	4	5	6 or more
Number of plants:	12	8	3	2	1	1	0

9. The following table gives the percentage classification of herds according to hematologic status in two locations.

	Hematologic status		
Locations	Negative	Suspected	Positive
I	80	8	12
II	63	20	17

Test the hypothesis of independence of locations and hematologic status of herds.

10. Test the goodness of fit of a normal distribution to the frequency distribution obtained in exercise 2, chapter 2.

11. In an experiment concerning the immunization of goats from anthrax, the following results were obtained. Derive an inference on the efficiency of the vaccine.

	Died	Survived
Inoculated	1	15
Not inoculated	8	9

12. In a botany experiment the results of crossing two hybrids of a species of flower gave observed frequencies of descriptive categories of 123, 47, 35, 10. Do these results agree with theoretical frequencies that specify a ratio of 9:3:3:1?

13. Analyze the following 2 × 2 contingency table to determine if the color in a certain variety of rabbits is sex related.

	Male	Female
White	2	6
Fawn	9	4

14. Two different fertilizers were applied to each of 50 plants to study the effect on plant height. The results are presented in the following table.

Fertilizer	Effect on growth	No effect on growth
I	39	11
II	22	28

Test the effect of the fertilizers.

15. The following table gives the results of a series of controlled experiments. Discuss whether the treatment has any positive effect.

	Positive	No effect	Negative
Treatment	15	4	3
Control	5	9	7

16. The normal infection rate for a certain disease in pigs is known to be 50%. In an experiment in which 100 pigs were injected with a new drug, it was found that 10 of the pigs contracted the infection. Can the evidence be regarded as conclusive (at the 5% level of significance) proof of the value of the new drug?

9. Correlation and Regression

In all the material presented so far, we have been concerned with the value of such single random variables as the weight of cans, the percentage of crop loss, or the length of cucumbers. We now consider the case where in the same experiment we view two variables at the same time to discover their degree of relationship. For example, we might wish to know how consumer preference for mayonnaise is affected by the amount of sugar added, or how the tenderness of turkeys is related to their age. If we take a random sample of n individuals from a population and for each we measure two variables X and Y, we will end up with a sequence of pairs of data as follows:

$$(X_1, Y_1), (X_2, Y_2), \ldots, (X_n, Y_n)$$

We shall consider two techniques for studying the relationship between X and Y. *Correlation* measures the degree of linear association between X and Y, while *regression* seeks to find an equation that mathematically expresses the relationship between X and Y.

9.1 Correlation

We assume that pairs of variables have some degree of association between them. Yearly amount of sunshine (X) and crop yield in bushels (Y) are usually strongly associated; they have a high positive linear correlation. The amount of growth hormone given to turkeys (X) and the number of weeks taken to reach maturity (Y) have a strong association, but in a negative sense, since the number of weeks to maturity decreases as the amount of hormone is increased; they have a high negative linear correlation. The yield of Virginia tobacco (X) and the yield of Saskatchewan wheat (Y) have no association; they have zero linear correlation.

For a population of individuals or subjects in which we are studying a pair of variables X and Y, it is assumed that there is some degree of linear correlation, denoted by ρ, which measures the strength of the association between X and Y. ρ is a dimensionless quantity between -1 and 1 inclusive. If ρ is -1 or 1, all values of (X,Y) for the population lie on a straight line. A positive value of ρ indicates that as X increases, Y increases too. A negative value of ρ indicates that as one variable increases the other decreases. If $\rho = 0$ we say that the variables are uncorrelated. Keep in mind that ρ measures only the linear relationship between X and Y. It is possible to have $\rho = 0$ but to have the two variables related in a quadratic sense.

ρ is a population parameter whose value we may wish to estimate or use to test hypotheses. To do this we take a random sample of pairs of observations (X_1,Y_1), (X_2,Y_2) , . . . (X_n,Y_n) and calculate the *sample correlation coefficient,* denoted r, where

$$r = \frac{\sum_{i=1}^{n} (X_i - \overline{X})(Y_i - \overline{Y})}{\sqrt{\sum_{i=1}^{n} (X_i - \overline{X})^2 \sum_{i=1}^{n} (Y_i - \overline{Y})^2}}$$

or using a computational formula

$$r = \frac{\Sigma X_i Y_i - n\overline{X}\,\overline{Y}}{\sqrt{[\Sigma X_i^2 - n\overline{X}^2][\Sigma Y_i^2 - n\overline{Y}^2]}}$$

It may be shown mathematically that r always lies between -1 and 1 inclusive.

Example 9.1 In a study of the percentage of growth hormone in the diet (X) and the weight at 20 weeks in kg of turkeys (Y), the following random sample was obtained. Calculate the correlation coefficient r.

Hormone percentage (X)	4	6	8	10	12	14	16	18	20
Weight (Y)	4.1	4.9	4.6	5.3	5.8	5.6	6.4	6.4	6.7

$\Sigma X = 108, \quad \Sigma Y = 49.8, \quad \overline{X} = 12, \quad \overline{Y} = 5.53, \quad n = 9$
$\Sigma X^2 = 1536, \Sigma Y^2 = 281.88, \Sigma XY = 635.2, r = 0.9654$

To use r to calculate confidence intervals for ρ or to use r to test hypotheses concerning ρ, we must assume that X and Y have a bivariate normal distribution. For all practical purposes, this is equivalent to saying that X has a normal distribution and Y has a normal distribution. This being so, if $\rho = 0$, the expression

$$t = \frac{r}{\sqrt{(1 - r^2)/(n - 2)}}$$

has approximately a t-distribution with $n - 2$ degrees of freedom. A question of some importance to an experimenter is whether X and Y are correlated at all. This is equivalent to testing H_0: $\rho = 0$ against H_A: $\rho \neq 0$. We reject at the α level of significance if $t < -t_{\alpha/2}$ or if $t > t_{\alpha/2}$. If the alternative hypothesis is H_A: $\rho < 0$, we reject H_0 if $t < -t_\alpha$. If the alternative hypothesis is H_A: $\rho > 0$, we reject H_0 if $t > t_\alpha$.

Example 9.2 Use the results of example 9.1 to test that $\rho = 0$, i.e., to test H_0: $\rho = 0$ against H_A: $\rho \neq 0$ at a 0.05 level of significance.

$$t = \frac{0.9654}{\sqrt{(1 - .9320)/7}} = 9.79$$

Since $t_{\alpha/2} = 2.36$, we reject the null hypothesis and conclude that $\rho \neq 0$.

If $\rho = \rho_0 \neq 0$, the expression

$$\frac{1}{2}\log_e\left(\frac{1 + r}{1 - r}\right)$$

has approximately a normal distribution with mean μ and standard deviation σ given as follows:

$$\mu = \frac{1}{2}\log_e\left(\frac{1 + \rho_0}{1 - \rho_0}\right), \quad \sigma = \frac{1}{\sqrt{n - 3}}$$

so that

$$z = \frac{\frac{1}{2}\log_e\left(\frac{1+r}{1-r}\right) - \frac{1}{2}\log_e\left(\frac{1+\rho_0}{1-\rho_0}\right)}{1/\sqrt{n-3}}$$

has a standard normal distribution. Thus, to test H_0: $\rho = \rho_0$, H_A: $\rho \neq \rho_0$ at level of significance α, we reject H_0 if $z < -Z_{\alpha/2}$ or if $z > Z_{\alpha/2}$. If the alternative hypothesis is H_A: $\rho < \rho_0$, we reject H_0 if $z < -Z_\alpha$. If the alternative hypothesis is H_A: $\rho > \rho_0$, we reject H_0 if $z > Z_\alpha$. Values of $\frac{1}{2}\log_e\left(\frac{1+r}{1-r}\right)$ and $\frac{1}{2}\log_e\left(\frac{1+\rho_0}{1-\rho_0}\right)$ are found in Appendix Table 9.

Example 9.3 Use the results of example 9.1 to test H_0: $\rho = 0.5$, H_A: $\rho > 0.5$ at a 0.05 level of significance.

$$\frac{1}{2}\log_e\left(\frac{1+r}{1-r}\right) = \frac{1}{2}\log_e\left(\frac{1+.97}{1-.97}\right) = 2.092$$

$$\frac{1}{2}\log_e\left(\frac{1+\rho_0}{1-\rho_0}\right) = \frac{1}{2}\log_e\left(\frac{1+.5}{1-.5}\right) = 0.549$$

$$z = (2.092 - 0.549)/0.4082 = 3.78$$

Since $z > Z_\alpha$ we reject H_0 and conclude that $\rho > 0.5$.

We know from chapter 7 that hypothesis testing and confidence intervals are related. A $100(1 - \alpha)\%$ confidence interval for μ is

$$\left(\frac{1}{2}\log_e\left(\frac{1+r}{1-r}\right) - \frac{Z_{\alpha/2}}{\sqrt{n-3}}, \frac{1}{2}\log_e\left(\frac{1+r}{1-r}\right) + \frac{Z_{\alpha/2}}{\sqrt{n-3}}\right)$$

and a $100(1 - \alpha)\%$ confidence interval for ρ can be read off from Appendix Table 9.

Example 9.4 Use the results of example 9.1 to calculate a 95% confidence interval for ρ.

A 95% confidence interval for μ is

$$\left(\frac{1}{2}\log_e\left(\frac{1+0.97}{1-0.97}\right) - \frac{1.96}{\sqrt{6}}, \frac{1}{2}\log_e\left(\frac{1+0.97}{1-0.97}\right) - \frac{1.96}{\sqrt{6}}\right)$$

or (1.292, 2.892).

From Appendix Table 9 we read off the corresponding confidence limits to be 0.86 and 0.99. (Note: the lower limit is greater than 0.99, but for the accuracy of the table, 0.99 will suffice. A more accurate value would be 0.994.)

9.2 Linear Regression

If we assume that the X and Y values have a bivariate normal distribution, we can calculate the correlation coefficient. If the correlation coefficient is significantly different from zero, the two variables are associated or interrelated. Note that there is no implication of causality (cause and effect) by a non-zero correlation coefficient.

Suppose now that we know or are willing to assume that the relationship between X and Y, if it exists at all, is linear. Even so, the observed points (X_1,Y_1), (X_2,Y_2), . . . , (X_n,Y_n) will not lie exactly on a straight line because of sampling variability, but instead will be scattered about the line. We assume that for each value of X there is a distribution of Y-values and that our Y-values for a particular value of X are a random sample from that distribution. The line that joins the means of all the distributions is called the *regression line.* Because a straight line is described by the equation $y = a + bx$, the equation of the regression line (which we shall refer to as the population regression line) is

$$\mu_{Y|X} = A + BX$$

where $\mu_{Y|X}$ is the mean value of Y for a given value of X. This is illustrated in figure 9.1. The constants A and B are called the *intercept* and *slope* of the line.

If the whole population of (X,Y) values were measured, A and B would be known. They are the parameters of the population. However, since, in general, the whole population cannot be measured, A and B must be estimated from the results of a random sample.

For any observed point (X_i,Y_i), $Y_i = \mu_{Y|X_i} + \varepsilon_i$ where ε_i is the random difference between the observed Y_i and the mean $\mu_{Y|X_i}$. This is illustrated in figure 9.2.

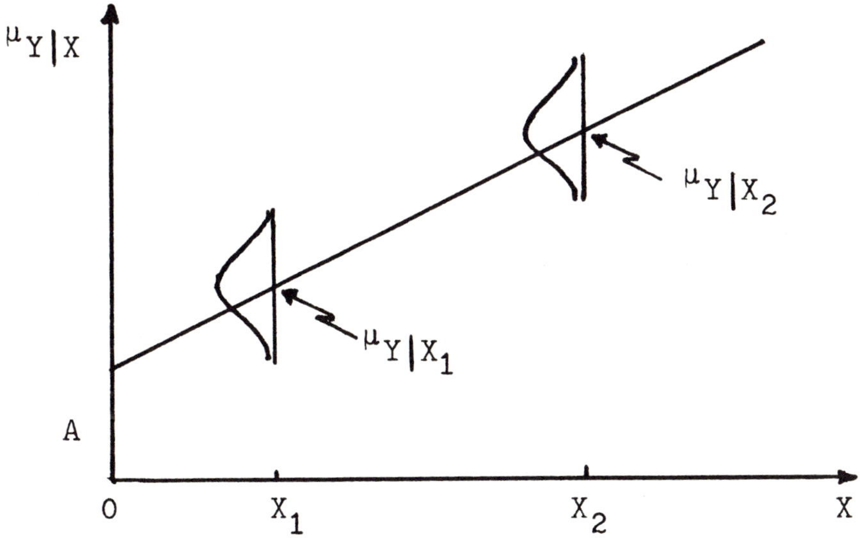

Fig. 9.1 *The mean of the distribution of Y for fixed X.*

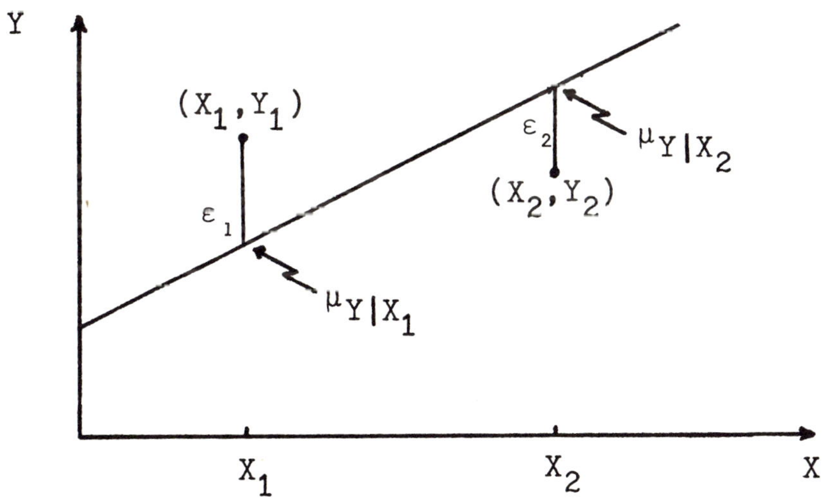

Fig. 9.2 *Relationships between* $\mu_{Y|X_i}$, Y_i *and* ε_i.

Replacing $\mu_{Y|X_i} = Y_i - \varepsilon_i$ in equation $\mu_{Y|X_i} = A + BX_i$,

$$\text{we have } Y_i = A + BX_i + \varepsilon_i$$

We estimate A and B by the *method of least squares,* whereby the estimates of A and B are calculated so that the sum of the squares of the ε-values is minimized. We denote the least square estimators of A and B by a and b. The estimator of the population regression equation is thus

$$\hat{Y} = a + bX$$

which we refer to as the estimated regression line where $\hat{Y}$, a, and b are respectively estimators of $\mu_{Y|X}$, A, and B. For the method of least squares to be valid the following three conditions should be met.

1. The sample $(X_1,Y_1), (X_2,Y_2), \ldots, (X_n,Y_n)$ is random, so that the differences $\varepsilon_1, \varepsilon_2, \ldots, \varepsilon_n$ can be considered uncorrelated.
2. For each value X_i of X, the variance of the distribution Y given X_i is the same, namely σ^2.
3. The X-values are known (i.e., X is not a random variable).

In addition, if we wish to calculate confidence intervals for A and B and test hypotheses concerning them, we will need to assume that the distribution of the Ys for each value X_i of X is normal with mean $\mu_{Y|X_i} = A + BX_i$ and variance σ^2.

The process of finding estimators of A and B that minimize the sum of the squares of the ε-values involves the use of advanced mathematics. Suffice it to say that we can derive the formulas for a and b given here.

$$b = \frac{\Sigma (X_i - \overline{X})(Y_i - \overline{Y})}{\Sigma (X_i - \overline{X})^2}, \quad a = \overline{Y} - b\overline{X}$$

A computationally simpler formula for b is

$$b = \frac{\Sigma X_iY_i - (\Sigma X_i)(\Sigma Y_i)/n}{\Sigma X_i - (\Sigma X_i)^2/n}$$

Example 9.5 Suppose that using the data of example 9.1 we assumed that the relationship between hormone level (X) and weight at 20 weeks (Y) was linear. Calculate a, b, and $\hat{Y}$ and draw the estimated regression line on the scatter diagram of X versus Y in figure 9.3.

$$b = \frac{635.2 - (108)(49.8)/a}{1356 - (108)^2/9} = 0.157$$

$$a = 5.53 - (0.157)(12) = 3.646$$

$$\hat{Y} = 3.646 + 0.157X$$

To draw the estimated regression line, find $\hat{Y}$ for two values of X (say 4 and 20). When $X = 4$, $\hat{Y} = 4.274$. When $X = 20$, $\hat{Y} = 6.786$. Join the two points with a straightedge.

9.2.1 Partitioning of the Sum of Squares

When we are dealing with a simple variable, say Y, such as weight of cans of fruit, the observations $Y_1, Y_2, \ldots, Y_n$ are assumed to

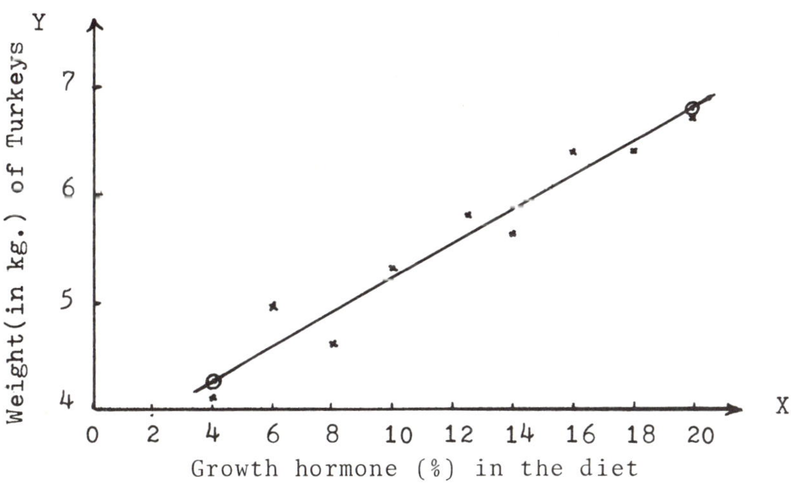

Fig. 9.3 *Scatter diagram and estimated regression line.*

come from a normal distribution with mean μ and variance σ^2. The variation of the Y-values is measured by the variance which, in turn, is estimated by S^2. The numerator of the formula for S^2 is $\Sigma(Y_i - \overline{Y})^2$ and called a *sum of squares.*

When dealing with a regression problem, some of the variation in the Y-values is due to the regression between X and Y (that is, X and Y are linearly related, so as X changes it causes changes in the Ys) and some is due to the variation of the Y-values about the regression line. We would like to separate these two sources of variation in Y.

Consider figure 9.4 in which a single observed point (X_i, Y_i) is shown.

We can see that

$$(Y_i - \overline{Y}) = (Y_i - \hat{Y}_i) + (\hat{Y}_i - \overline{Y})$$

From figure 9.4, $Y_i - \hat{Y}_i$ is due to the variation of Y_i from the regression line, and $\hat{Y}_i - \overline{Y}$ is due to the regression line itself. If we sum both sides of this expression for all observations from 1 to

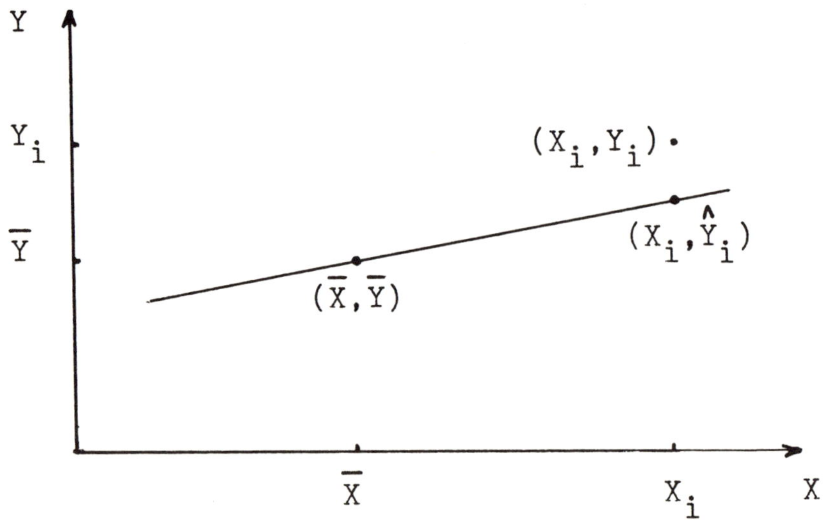

Fig. 9.4 *Sources of variation of Y's.*

n and then square the sums, we can show (see exercise 9.7) that

$$\Sigma\,(Y_i - \overline{Y})^2 = \Sigma\,(Y_i - \hat{Y}_i)^2 + \Sigma\,(\hat{Y}_i - \overline{Y})^2$$

The left expression is the total sum of squares, denoted SST. The central expression is the error sum of squares (SSE) and measures the variation about the regression line. The right expression is called the regression sum of squares, denoted SSR. A simpler way to calculate SSE and SSR using already calculated quantities is

$$SSE = \Sigma\,(Y_i - \overline{Y})^2 - b\,\Sigma\,(X_i - \overline{X})(Y_i - \overline{Y})$$

$$SSR = b\,\Sigma\,(X_i - \overline{X})(Y_i - \overline{Y})$$

The degree of freedom associated with SST is n − 1 because, to calculate it from the n observations, we first calculate $\overline{Y}$, therefore losing a degree of freedom. The degree of freedom associated with SSE is n − 2 because, from the n observations, we must first calculate the regression line, which has two parameters (A and B) and therefore loses two degrees of freedom. The degree of freedom associated with SSR is 1 because, of the two degrees of freedom allowed for $\hat{Y}_i$, we must first calculate $\overline{Y}$, therefore losing one degree of freedom. The sums of squares and degrees of freedom are summarized in table 9.1.

Table 9.1 *Partitioning sum of squares in regression.*

Source of variation	Degrees of freedom	Sum of squares
Due to regression	1	$SSR = \Sigma\,(\hat{Y}_i - \overline{Y})^2$
Deviations from regression (Error)	n − 2	$SSE = \Sigma\,(Y_i - \hat{Y}_i)^2$
Total	n − 1	$SST = \Sigma\,(Y_i - \overline{Y})^2$

It is of some interest to look at the square of the correlation coefficient in light of this partitioning of the sum of squares:

$$r^2 = \frac{[\Sigma\,(X_i - \overline{X})(Y_i - \overline{Y})]^2}{\Sigma\,(X_i - \overline{X})^2\,\Sigma\,(Y_i - \overline{Y})^2}$$

$$= \frac{b\,\Sigma\,(X_i - \overline{X})(Y_i - \overline{Y})}{\Sigma\,(Y_i - \overline{Y})^2}$$

The numerator is the sum of squares due to regression and the denominator is the total sum of squares. Thus, r^2 measures the proportion of the total sum of squares due to regression and $1 - r^2$ expresses the proportion of the total sum of squares due to variation about the regression line. r^2 itself is usually referred to as the *coefficient of determination.* Notice that r^2 will always be between 0 and 1, which verifies the statement made earlier that $-1 \leq r \leq 1$.

Example 9.6 Partition the sum of squares for the data of example 9.1

$$SSR = b\,\Sigma\,(X_i - \overline{X})(Y_i - \overline{Y}) = b[\Sigma\,XY - (\Sigma X)(\Sigma Y)/n]$$

$$= (0.157)[635.2 - (108)(49.8)/9] = 5.903$$

$$SSE = \Sigma\,(Y_i - \overline{Y})^2 - SSR = \Sigma\,Y_i^2 - (\Sigma Y_i)^2/n - SSR$$

$$= 6.320 - 5.903 = 0.417$$

$$SST = \Sigma\,(Y_i - \overline{Y})^2 = 6.320$$

9.2.2 Variance Estimates

The variance of Y given X, σ^2, measures the dispersion of Y-values about the regression line. Calling $S^2_{Y|X}$ the estimator of that variance,

$$S^2_{Y|X} = \frac{\Sigma\,(Y_i - \hat{Y}_i)^2}{n - 2} = \frac{SSE}{n - 2}$$

The square root of this expression, the *standard error,* is

$$S_{Y|X} = \sqrt{\frac{SSE}{n-2}}$$

Since a and b are estimators of A and B, they too are subject to sampling variability. It can be shown that (i) the distribution of a has mean A and variance σ_a^2 where

$$\sigma_a^2 = \sigma^2 \left(\frac{\Sigma X_i^2}{n \, \Sigma (X_i - \overline{X})^2} \right)$$

and (ii) the distribution of b has mean B and variance σ_b^2 where

$$\sigma_b^2 = \sigma^2 / \Sigma (X_i - \overline{X})^2$$

so that estimators of σ_a^2 and σ_b^2 are S_a^2 and S_b^2 where

$$S_a^2 = S_{Y|X}^2 \left(\frac{\Sigma X_i^2}{n \, \Sigma (X_i - \overline{X})^2} \right) \qquad S_b^2 = S_{Y|X}^2 / \Sigma (X_i - \overline{X})^2$$

Example 9.7 Calculate $S_{Y|X}^2$, S_a^2, and S_b^2 for the data of example 9.1.

$$S_{Y|X}^2 = \frac{SSE}{n-2} = \frac{0.417}{7} = 0.0596$$

$$S_a^2 = (0.0596) \left(\frac{1536}{(9)(240)} \right) = 0.0424$$

$$S_b^2 = (0.0596)/240 = 0.0002$$

9.2.3 Confidence Intervals and Tests of Hypothesis for A and B

All of the results, this far, depend only on the assumptions listed in section 9.2, namely, that the Y-values are uncorrelated and the distribution of Y given X_i has the same variance σ^2.

If, in addition, we assume that the distribution of the Y-values is normal, we will be able to test hypotheses and calculate confidence intervals for A and B. We can show that, if the Y-values have a normal distribution, a and b will also have normal distributions, so that

$$t_1 = \frac{b - B}{S_b} \quad \text{and} \quad t_2 = \frac{a - A}{S_a}$$

have t-distributions with n − 2 degrees of freedom because S_a and S_b depend on $S_{Y|X}$ which has n − 2 degrees of freedom. A 100(1 − α)% confidence interval for A is

$$(a - t_{\alpha/2}\, S_a,\ a + t_{\alpha/2}\, S_a)$$

A 100(1 − α)% confidence interval for B is

$$(b - t_{\alpha/2}\, S_b,\ b + t_{\alpha/2}\, S_b)$$

Similarly, to test H_0: B = B_0, H_A: B ≠ B_0 at significance level α, we reject H_0 if $t_1 < -t_{\alpha/2}$ or $t_1 > t_{\alpha/2}$. If the alternative hypothesis is H_A: B < B_0, we reject H_0 if $t_1 < t_\alpha$. If the alternative hypothesis is H_A: B > B_0, we reject H_0 if $t_1 > t_\alpha$. Parallel results hold for testing A.

Example 9.8 Using the data of example 9.1 calculate a 95% confidence interval for A and test H_0: B = 0, H_A: B ≠ 0. A 95% confidence interval for A is

$$(a - t_{\alpha/2}\, S_a,\ a + t_{\alpha/2}\, S_a)$$

or [3.646 − (2.36)(0.2059), 3.646 + (2.36)(0.2059)]
or (3.16, 4.13).

To test the hypothesis for B, the test statistic is

$$t_1 = \frac{b - B_0}{S_b} = \frac{0.157 - 0}{0.0158} = 9.937$$

Since $t_1 > t_{\alpha/2}$, we reject H_0 and conclude that B ≠ 0.

9.2.4 X and Y Both Random Variables

In the discussion so far we assumed that the X-values were not random variables but were known or set by the experimenter. This enabled us to use the methods of least squares to estimate A and B. If we must assume both that X and Y are random variables, and that X and Y have a bivariate normal distribution, a different method of estimation, the *maximum likelihood method* is used. This yields estimators of A and B exactly the same as the least squares estimators. The formulas for estimating A and B may be used whether X is considered as known or as a random variable.

9.2.5 Using $\hat{Y}$ as a Predictor

Two problems of interest, which can be solved once a, b, and $\hat{Y}$ have been calculated, are the following:

1. For a particular value of X, say X_0, what is the mean value of Y, i.e., what is $\mu_{Y|X_0}$?
2. For a particular value of X, say X_0, what is the predicted value of Y?

To answer the first question, consider $\hat{Y} = a + bX$, and suppose that $\hat{Y}_0$ is the value of $\hat{Y}$ when $X = X_0$, so that

$$\hat{Y}_0 = a + bX_0$$

It can be shown that $\hat{Y}_0$ has a normal distribution with mean $\mu_{Y|X_0}$ and variance

$$\sigma^2 \left(\frac{1}{n} + \frac{(X_0 - \overline{X})^2}{\Sigma (X_i - \overline{X})^2} \right)$$

so that $\hat{Y}_0$ is a point estimator of $\mu_{Y|X_0}$. A $100(1 - \alpha)\%$ confidence interval for $\mu_{Y|X_0}$ is

$$(\hat{Y}_0 - t_{\alpha/2}\, S_{\hat{Y}_0},\ \hat{Y}_0 + t_{\alpha/2}\, S_{\hat{Y}_0})$$

where

$$S^2_{\hat{Y}_0} = \left(\frac{1}{n} + \frac{(X_0 - \overline{X})^2}{\Sigma\,(X_i - \overline{X})^2}\right) S^2_{Y|X}$$

and t is found in the t-table with n − 2 degrees of freedom. Consideration of the confidence interval shows that the length of the interval depends on X_0, and that the interval length is narrowest for $X_0 = \overline{X}$. Further, as X_0 moves away from $\overline{X}$, the interval becomes wider, so that if X_0 is too far from $\overline{X}$, the interval is impractically wide. This indicates that one should not generally use regression equations to predict $\mu_{Y|X_0}$ if X_0 is very far from $\overline{X}$.

To find the predicted value of Y for a particular value X_0 of X, we use the fact, which can be shown by the use of mathematical statistics, that a $100(1 - \alpha)\%$ *prediction interval* is

$$(\hat{Y}_0 - t_{\alpha/2}\, S_{Y_0},\ \hat{Y}_0 + t_{\alpha/2}\, S_{Y_0})$$

where

$$S^2_{Y_0} = \left(1 + \frac{1}{n} + \frac{(X_0 - \overline{X})^2}{\Sigma\,(X_i - \overline{X})^2}\right) S^2_{Y|X}$$

Example 9.9 Calculate a 95% confidence interval for $\mu_{Y|X}$ when X = 6. Then calculate a 95% prediction interval for Y when x = 15. A 95% confidence interval for $\mu_{Y|X}$ is

$$(\hat{Y}_0 - t_{\alpha/2}\, S_{\hat{Y}_0},\ \hat{Y}_0 + t_{\alpha/2}\, S_{\hat{Y}_0})$$

where $\hat{Y}_0 = 3.646 + (0.157)(6) = 4.588$

$$S^2_{\hat{Y}_0} = \left(\frac{1}{9} + \frac{36}{240}\right)(0.0596) = 0.0156$$

$$t_{\alpha/2} = 2.36$$

The required confidence interval is (4.29, 4.88).

A 95% prediction interval for Y when X = 15 is

$$(\hat{Y}_0 - t_{\alpha/2}\, S_{Y_0},\ \hat{Y}_0 + t_{\alpha/2}\, S_{Y_0})$$

where $S^2_{Y_0} = \left(1 + \frac{1}{9} + \frac{36}{240}\right)(0.0596) = 0.0752$

The required interval is

$$(6.001 - .647,\ 6.001 + .647)$$

or $(5.354, 6.648)$

Exercises

1. The following figures show results of an experiment where weights of a food additive (Y) were dissolved in 100 gm of water at different temperatures (X).

X	0	10	20	30	40	50	60
Y	55	59	65	70	75	81	86

   Assuming that X and Y have a bivariate normal distribution,
   (a) Calculate the correlation coefficient r.
   (b) Calculate the coefficient of determination.
   (c) Test that $\rho = 0$ and that $\rho = 0.5$ (use $\alpha = 0.05$).
   (d) Set a 90% confidence interval for ρ.

2. The following figures show the results of an experiment where the freshness of vegetables (judged by a panel of experts) on a ten-point scale (Y) was related to the time from picking (X) in days.

X	0	1	2	3	4	5	6	7	8	9	10
Y	9.7	9.4	8.6	7.0	6.2	6.8	5.6	4.3	4.0	2.7	1.8

   Assuming that X and Y have a bivariate normal distribution,
   (a) Calculate the correlation coefficient r.
   (b) Calculate the coefficient of determination.
   (c) Test that $\rho = 0$ and that $\rho = -0.8$ (use $\alpha = 0.1$).
   (d) Set a 95% confidence interval for ρ.

3. In a study of the reaction rate (Y) in mg/unit time against enzyme concentration (X) in ml, the following figures were obtained

X	1.0	1.5	2.0	2.5	3.0	3.5	4.0
Y	105	193	252	298	387	450	510

Assuming that X and Y have a bivariate normal distribution,
   (a) Calculate the coefficient of determination and correlation coefficient.
   (b) Test that $\rho = 0$ and that $\rho = 0.8$ (use $\alpha = 0.05$).
   (c) Set a 90% confidence interval for ρ.

4. In exercise 1 suppose that we can assume that Y is affected by X in a linear manner. Using linear regression,
   (a) Estimate the regression line $\mu_{Y|X} = A + BX$.
   (b) Partition the sum of squares and obtain $S^2_{Y|X}$ and the standard error.
   (c) Assuming that the distribution of Y for a fixed X is normal, set 95% confidence intervals for A and B.
   (d) Test (with $\alpha = 0.05$) H_0: $B = 1$, H_A: $B > 1$.
   (e) Calculate a 95% confidence interval for $\mu_{Y|X}$ when $X = 35$.

5. In exercise 2 suppose that we can assume that Y has a normal distribution with mean $\mu_{Y|X}$, and that $\mu_{Y|X} = A + BX$.
   (a) Estimate the regression line $\mu_{Y|X} = A + BX$.
   (b) Partition the sum of squares, calculate $S^2_{Y|X}$, and the standard error.
   (c) Calculate 90% confidence intervals for A and B.
   (d) Test (with $\alpha = 0.10$) H_0: $B = 1$, H_A: $B \neq 1$, and also H_0: $A = 10$, H_A: $A < 10$.
   (e) Calculate a 95% confidence interval for $\mu_{Y|X}$ when $X = 6$.

6. In exercise 3 suppose that we can assume that Y has a normal distribution with mean $\mu_{Y|X}$ and that $\mu_{Y|X} = A + BX$.
   (a) Estimate the regression line $\mu_{Y|X} = A + BX$.
   (b) Partition the sum of squares, calculate $S^2_{Y|X}$, and the standard error.

(c) Calculate 90% confidence intervals for A and B.
(d) Test (with $\alpha = 0.05$) H_0: B = 100, H_A: B > 100, and H_0: A = 0, H_A: A ≠ 0.
(e) Calculate a 95% confidence interval for $\mu_{Y|X}$ when X = 1.8.

7. Show for the linear regression case that $\Sigma(Y_i - \overline{Y})^2 = \Sigma(Y_i - \hat{Y}_i)^2 + \Sigma(\hat{Y}_i - \overline{Y})^2$ by showing that the crossproduct term is 0.

10.

Analysis of Variance

In chapter 7 we learned how to test hypotheses concerning the means of two populations. It is now time to consider what to do in the case of more than two populations. One seemingly reasonable approach is to test the means two at a time until all pairs of comparisons have been made. However, this is not a fruitful method for the two following reasons: (i) The number of pairs of tests becomes burdensome as the number of populations increases. For instance, for 10 populations there are 45 pairs of tests necessary, and (ii) the null hypothesis of no differences between the population means would be rejected due to sampling variation, not real differences, with probability α (using each test at significance level α). If we recall from chapter 4 that we interpret probability as a relative frequency, this last statement means that about 100 α% of the tests will reject H_0 even if H_0 is true. Therefore, even if there are no differences between the population means, about 100α% of the tests will indicate a difference and reject the null hypothesis.

A beginner might wonder if, in comparing k means, we can avoid the battery of t-tests by comparing the largest and the smallest means. The answer to this is emphatically, "No."

Table 10.1 shows the consequences of comparing the extremes (largest against smallest) of k means even when all the means are from the same distribution.

Table 10.1 *Consequences of comparing extremes of k means.*

Number of means being compared	Probability with which the extremes will be significantly different, using α significance level tests
2	0.05
3	0.13
6	0.40
20	0.90

This table shows that if an experimenter compares two means which *appear* to be different, the chance of making type I errors (rejecting H_0 even when H_0 is true) can be very large. The need to avoid such a procedure cannot be overemphasized.

One of the most important techniques for dealing with the problem of comparing the means of more than two populations is called *analysis of variance.* It is one of the most useful tools available to a statistician and involves partitioning the total variation in an experiment according to the sources of variation in the experiment. We now consider some basic experimental designs that use variance analysis.

10.1 Completely Randomized Design

Suppose that there are k treatments being compared, each of which is to be applied to n subjects for a total of kn subjects. Each treatment is to be applied to the subjects randomly as follows: For each subject a random number between 1 and k is selected to determine which treatment it will receive. The random numbers can be numbers drawn from a hat or found by tossing a coin or more simply by using a table of random numbers. We usually impose on the randomization scheme the restriction that each treatment number will only be used n times. An experiment where treatments are assigned in this way is called a *completely randomized design* (CRD). Each subject has an equal chance of receiving any treatment. At the end of the experiment there are kn observations in k groups each of size n. The n observations in any group are regarded as a random sample from the population of all responses for that treatment. We assume that the variation among the observations with one treatment is the same as the variation among observations within any other treatment and that each observation has a normal distribution. More simply said, we assume that the observations are k independent random samples from normal distributions, each distribution having the same variance. To partition the variation into its various components, it will be convenient to adhere to the following notation:

k = number of populations (and number of samples)
n = number of observations per sample
X_{ij} = *j*th observation in the *i*th sample
T_i = sum of observations in sample i
$\overline{X}_i$ = mean of observations in sample i
T = sum of all observations
$\overline{X}$ = mean of all observations (grand mean)

The data may be displayed as follows in table 10.2.

Table 10.2 *Completely randomized design data.*

Sample 1	Sample 2	...	Sample k
X_{11}	X_{21}		X_{k1}
X_{12}	X_{22}		X_{k2}
.	.		.
.	.		.
.	.		.
X_{1n}	X_{2n}		X_{kn}
T_1	T_2	...	T_k
$\overline{X}_1$	$\overline{X}_2$	...	$\overline{X}_k$

For the *i*th sample, $\overline{X}_i = \sum_{j=1}^{n} X_{ij}/n$

and $$\overline{X} = \sum_{i=1}^{k} \sum_{j=1}^{n} X_{ij}/nk$$

10.1.1 Partitioning of the Sum of Squares

Total variation in the experiment is measured by the variance of all observations about the grand mean, i.e.,

$$\frac{\sum_{i=1}^{k}\sum_{j=1}^{n}(X_{ij} - \overline{X})^2}{nk - 1}$$

Now consider the deviation $X_{ij} - \overline{X}$

$$X_{ij} - \overline{X} = (X_{ij} - \overline{X}_i) + (\overline{X}_i - \overline{X})$$

Squaring both sides and summing gives

$$\sum_{i=1}^{k}\sum_{j=1}^{n}(X_{ij} - \overline{X})^2 = \sum_{i=1}^{k}\sum_{j=1}^{n}(X_{ij} - \overline{X})^2 +$$

$$\sum_{i=1}^{k} n(\overline{X}_i - \overline{X})^2 + 2\sum_{i=1}^{k}(\overline{X}_i - \overline{X})\sum_{j=1}^{n}(X_{ij} - \overline{X}_i)$$

The last term is 0, since $\Sigma\,(X_{ij} - \overline{X}_i)$ is the sum of the observations of the *i*th sample from their mean. Notice that the left-hand expression is the numerator when calculating the variance of all observations about the grand mean.

$\sum_{i=1}^{k}\sum_{j=1}^{n}(X_{ij} - \overline{X})^2$ is called the total sum of squares (SST)

$\sum_{i=1}^{k} n(\overline{X}_i - \overline{X})^2$ is called the treatment sum of squares (SSTR)

$\sum_{i=1}^{k}\sum_{j=1}^{n}(X_{ij} - \overline{X}_i)^2$ is called the error sum of squares (SSE)

The treatment sum of squares (SSTR) measures the variation of treatment means about the grand mean and (SSE) measures the variation of observations about their own treatment mean. This result is significant because it means that the total variation in an experiment (as measured by the total sum of squares or SST) can be partitioned into two components. The first of these, the SSTR, measures the variation of treatment means about the grand mean and thus measures difference between treatments. The second, the SSE, measures the difference between observations having the same treatment. It thus measures the experimental or inherent variation in an experiment. As we shall see, the relative sizes of these two sources of variation allows us to say whether a treatment's effect is significant or not.

Computationally more convenient formulas for calculating SST, SSTR, and SSE are given as follows:

$$SST = \sum_{i=1}^{k} \sum_{j=1}^{n} X_{ij}^2 - \frac{T^2}{nk}$$

$$SSTR = \sum_{i=1}^{k} \left(\frac{T_i^2}{n}\right) - \frac{T^2}{nk}$$

$$SSE = \sum_{i=1}^{k} \sum_{j=1}^{n} X_{ij}^2 - \sum_{i=1}^{k} \left(\frac{T_i^2}{n}\right)$$

Example 10.1 Three chemical formulations are to be compared on the basis of their ability to inhibit the ripening of tomatoes in transit. Twelve tomatoes are to be used and the three treatments are assigned to the tomatoes at random so that each treatment is used on four tomatoes. The response is the number of days to ripening. Partition the total sum of squares into that part due to treatments and that part due to experimental (error) variation.

Chemical 1	Chemical 2	Chemical 3	
12	15	6	
10	12	11	
9	13	10	
9	12	7	
$T_1 = 40$	$T_2 = 52$	$T_3 = 34$	$T = 126$

$$SST = 12^2 + 10^2 + \ldots + 7^2 - \frac{(126)^2}{12} = 71$$

$$SSTR = \frac{40^2 + 52^2 + 34^2}{4} - \frac{(126)^2}{12} = 42$$

$$SSE = 12^2 + 10^2 + \ldots + 7^2 - \left(\frac{40^2 + 52^2 + 34^2}{4}\right) = 29$$

Note that SST = SSTR + SSE

10.1.2 Construction of the Analysis of Variance Table

We saw in previous chapters that to estimate the variance from a sum of squares, we need to divide by the appropriate degrees of freedom. For the SST there are kn observations, but since $\overline{X}$ must first be calculated, the degrees of freedom are kn − 1. For the SSTR there are k values (of $\overline{X}_i$) to be considered, but since $\overline{X}$ must first be calculated, the degrees of freedom are k − 1. For the SSE there are kn values (of X_{ij}) to be considered, but we must first calculate k values (of $\overline{X}_i$), giving kn − k degrees of freedom. A sum of squares divided by its degrees of freedom is referred to as a *mean square*.

$\frac{SST}{kn - 1}$ is called the total mean square (MST)

$\frac{SSTR}{k - 1}$ is called the treatment mean square (MSTR)

$\frac{SSE}{k(n - 1)}$ is called the error mean square (MSE)

The MST is rarely used. The MSE estimates the within-sample variance and the MSTR estimates the between (or among) treatment variances. It is usual to present the sums of squares, degrees of freedom, and mean squares in an *analysis of variance table* (ANOVA table) like that shown in table 10.3.

Table 10.3 *Analysis of variance (CRD).*

Source of variation	Sum of squares	Degrees of freedom	Mean squares
Between treatments	SSTR	$k - 1$	MSTR
Within groups	SSE	$k(n - 1)$	MSE
Total	SST	$kn - 1$	MST

Under the heading "Source of variation" we sometimes abbreviate to "Between" and "Within."

Example 10.2 Continue with the data of example 10.1 and work out the ANOVA table. The degrees of freedom for total, treatments, and error are, respectively, 11, 2, and 9. The mean squares are as follows:

$$\text{MST} = 71/11 = 6.45$$
$$\text{MSTR} = 42/2 = 21.0$$
$$\text{MSE} = 29/9 = 3.22$$

Source of variation	Sum of squares	Degrees of freedom	Mean square
Between chemicals	42.0	2	21.00
Within groups	29.0	9	3.22
Total	71.0	11	6.45

10.1.3 The Mathematical Model

It is convenient at this time to introduce the idea of a mathematical model for the experiment. We suppose that each population from which a sample is drawn has mean μ_i ($i = 1, 2, \ldots, k$). Each observation can be thought of as consisting of the population mean from which it is drawn plus a random error ε_{ij}. Thus X_i is a normal random variable with mean μ_i, and we can write

$$X_{ij} = \mu_i + \varepsilon_{ij}$$

If we write

$$\mu = \frac{1}{k}\sum_{i=1}^{k} \mu_i$$

and

$$\tau_i = \mu_i - \mu$$

we can alternatively write

$$X_{ij} = \mu + \tau_i + \varepsilon_{ij} \text{ for } i = 1, 2, \ldots, k \text{ and } j = 1, 2, \ldots, n$$

This last expression is called the *mathematical model* of the experiment. μ is sometimes called the overall mean and τ_i is the deviation of the *i*th population mean μ_i from the overall mean. Because the μ_i's are parameters, μ is assumed to be a parameter and the ε_{ij}'s are assumed to be random variables having a normal distribution with mean zero and common variance σ^2.

If the τ_i's are assumed to be unknown constants or parameters, we talk of the model as being a *fixed effects model,* or sometimes a *model I.* We generally assume a fixed effects model if the populations from which the samples were drawn constitute all the population of interest. On the other hand, if the populations are themselves samples from a larger population, we assume that the τ_i's are random variables. More specifically, if we assume that the τ_i's are random variables from a normal population with mean zero and variance σ_τ^2, we speak of the model as being a *random effects model* or *model II.* Whether the model is fixed or random does not affect the sums of squares, degrees of freedom, or mean squares, but it does affect the interpretation of the analysis of variance results, as we shall see.

An example of a fixed effects model is an experiment to compare three canning processes. The variable of interest might be the spoilage rate. Since the three processes constitute all processes of interest, the model would be

$$X_{ij} = \mu + \tau_i + \varepsilon_{ij}$$

where X_{ij} = *j*th observation on the *i*th process (i = 1, 2, 3)
μ = overall mean response
τ_i = deviation of the *i*th process from μ
ε_{ij} = random variation of the *j*th observation on the *i*th process

An example of a random effects model is an experiment in which the subject is the egg weight of white leghorn chickens. Ten chickens are chosen from a large population and their eggs weighed for a period of four weeks. The possible number of chickens in the population is large, so a sample of 10 was chosen on which to experiment. If we assume that the differential (genetic) effect of any chicken has a normal distribution, the random effects model is appropriate.

We will only consider fixed effects models unless otherwise stated.

10.1.4 Tests of Hypothesis (CRD)

To test the hypothesis that all the k population means μ_i are equal (or equivalently, that all the τ_i are equal), we consider the ratio of the MSTR and the MSE. This ratio is denoted by the letter F in honor of R.A. Fisher who originally developed the analysis of variance technique. Thus,

$$F = MSTR/MSE$$

Remembering that MSTR estimates the variance between treatment means and MSE estimates the variation within treatment groups, we see that the F-ratio is the ratio of two variances. Since F is a statistic computed from a random sample, it has a distribution, called the F-distribution. The F-distribution is the distribution of the ratio S_1^2/S_2^2 of two independent sample variances each from normal distributions having common variance σ^2. The concept of the distribution of S_1^2/S_2^2 is no different from the concept that the t-distribution is the distribution of all values of $(\overline{X} - \mu)\sqrt{n}/\sigma$ from a normal population or that the χ^2-distribution is the distribution of all values of $(n - 1)S^2/\sigma^2$ from a normal population. The F-distribution is tabulated in Appendix Tables 10.1 and 10.2.

If the null hypothesis of equal treatment means is true, the variance between treatment means will be the same as the variance within the treatment groups and the ratio MSTR/MSE will be close to 1. On the other hand, if the null hypothesis is not true, so that at least one mean μ_i is different from the others, the ratio MSTR/MSE will be greater than 1. We can thus test the hypothesis of equality of population means by considering the F-ratio. In fact, the null hypothesis is rejected at the α level of significance if $F > F_\alpha$ where F_α is the upper tail of the F-distribution (from Appendix Table 10). To use Appendix Table 10, we must specify the degrees of freedom f_1 for MSTR and the degrees of freedom f_2 for MSE. A second fruitful way of understanding the F-ratio is to reconsider the mathematical model. We will not be able to derive the results in this book, but instead will simply present them. We recall from chapter 7 that if we use an estimator g to estimate a parameter γ, that g, being a random variable, will have a distribution. This distribution will have a mean μ_g and a variance σ_g^2. If the mean of the distribution of g, μ_g, is equal to γ, we say that g is an unbiased estimator of γ. In general, we like estimators to be unbiased. It can be shown that if the model is

$$X_{ij} = \mu + \tau_i + \varepsilon_{ij} \text{ for } i = 1, 2, \ldots, k \text{ and } j = 1, 2, \ldots, n$$

where $\varepsilon_{ij} \sim N(0, \sigma^2)$ and the τ_i's are unknown constants, MSE is an unbiased estimator of σ^2 and MSTR is an unbiased estimator of

$$\sigma^2 + \frac{n}{k-1} \sum_{i=1}^{k} \tau_i^2$$

If all the means μ_i are equal, then $\tau_i = 0$ and MSTR is also an unbiased estimator of σ^2 and MSTR/MSE will be close to 1. If, on the other hand, at least one μ_i is different from the others, at least one τ_i will be non-0 and MSTR will be an estimator of a quantity greater than σ^2. Thus, the ratio MSTR/MSE will be greater than 1. If experimenters find MSTR/MSE to be "large," (i.e., larger than F_α) they will conclude that at least one μ_i is different from the others.

Example 10.3 Use the data of example 10.1 analyzed in example 10.2 to test the null hypothesis at significance level 0.05 that the three chemicals act equally in their ability to inhibit ripening.

The value of F is 21/3.22 = 6.56. Since the value of $F_{.05}$ for $f_1 = 2$ and $f_2 = 9$ is 4.26, we reject the null hypothesis and conclude that the three chemicals do not act equally to inhibit ripening.

10.1.5 Estimation of Treatment Effects

The variance of an individual observation X_{ij} is σ^2, which is estimated by MSE. The variance of $\overline{X}_i$ is σ^2/n, which is estimated by MSE/n. The treatment mean $\overline{X}_i$ may be regarded as the point estimator of μ_i, and a $100(1 - \alpha)\%$ interval for μ_i is

$$\left(\overline{X}_i - t_{\alpha/2}\sqrt{\frac{MSE}{n}}, \overline{X}_i + t_{\alpha/2}\sqrt{\frac{MSE}{n}}\right)$$

where $t_{\alpha/2}$ is read in Appendix Table 4 with $k(n - 1)$ degrees of freedom. Since the point estimator of $\mu_i - \mu_j$ is $\overline{X}_i - \overline{X}_j$, which has variance $2\sigma^2/n$ and is therefore estimated by 2MSE/n, a $100(1 - \alpha)\%$ confidence interval for $\mu_i - \mu_j$ is

$$\left(\overline{X}_i - \overline{X}_j - t_{\alpha/2}\sqrt{\frac{2MSE}{n}}, \overline{X}_i - \overline{X}_j + t_{\alpha/2}\sqrt{\frac{2MSE}{n}}\right)$$

Example 10.4 Use the data of example 10.1 to calculate a 95% confidence interval for μ_1 and a 90% confidence interval for $\mu_2 - \mu_3$.

$\overline{X}_1 = 10, \overline{X}_2 = 13, \overline{X}_3 = 8.5, t_{.025} = 2.26, t_{.05} = 1.83,$
$MSE = 3.22, n = 4$

A 95% confidence interval for μ_1 is

$$(10 - (2.26)\sqrt{0.805}, 10 + (2.26)\sqrt{0.805})$$

or (7.97, 12.03).
A 90% confidence interval for $\mu_2 - \mu_3$ is

$$(4.5 - (1.83)\sqrt{1.61}, 4.5 + (1.83)\sqrt{1.81})$$

or (2.18, 6.82).

10.1.6 Orthogonal Contrasts

If the null hypothesis is rejected, we conclude that at least one of the treatment means is different from the others. The next step is to find which means are different from each other and which are not different. One method for doing this, which is particularly useful if one of the treatments is a control or standard treatment against which the others are to be compared, is called the method of *orthogonal contrasts.*

If for the k treatments (k samples), each with n observations, $\overline{X}_1$, $\overline{X}_2, \ldots, \overline{X}_k$ are the same means, and if $a_1, a_2, \ldots, a_k$ are any set of the k constants, then

$$a_1\overline{X}_1 + a_2\overline{X}_2 + \ldots + a_k\overline{X}_k \text{ (or } \Sigma\, a_i\overline{X}_i)$$

is called a *linear combination* of the sample means. Of particular importance is the case where the values of $a_1 + a_2 + \ldots + a_k$ add up to 0, in which case the linear combination is called a *contrast.* Suppose that treatment 1 is a standard treatment and we wish to compare μ_1 against the means of μ_2, μ_3, and μ_4 (i.e., suppose that $k = 4$). We may do this by testing the hypothesis H_0: $\mu_1 = ⅓(\mu_2 + \mu_3 + \mu_4)$ or, equivalently, H_0: $3\mu_1 = \mu_2 + \mu_3 + \mu_4$, or, even better, H_0: $3\mu_1 - \mu_2 - \mu_3 - \mu_4 = 0$. As a test statistic we may use

$$3\overline{X}_1 - \overline{X}_2 - \overline{X}_3 - \overline{X}_4$$

which is a contrast with $a_1 = 3$, $a_2 = a_3 = a_4 = -1$. If C_1 is any contrast, the sum of squares associated with the contrast is

$$SS(C_1) = \frac{n(\Sigma\, a_i\overline{X}_i)^2}{\Sigma\, a_i^2}$$

which has one degree of freedom. A contrast is tested for significance by the F-ratio:

$$F = SS(C_1)/MSE$$

which is compared to the F-distribution with $f_1 = 1$ and $f_2 = k(n - 1)$ degrees of freedom. The null hypothesis is rejected at the α level of significance if $F > F_\alpha$.

Two contrasts $C_1 = \Sigma\, a_i\overline{X}_i$ and $C_2 = \Sigma b_i\overline{X}_i$ are said to be orthogonal if $\Sigma a_i b_i = 0$, and a set of contrasts are said to be *mutually orthogonal* if all pairs of contrasts in the set are orthogonal. It can be proven that for k means, there are $k - 1$ sets of mutually orthogonal contrasts. However, the sets are not unique. If a set of orthogonal contrasts is found with each contrast having one degree of freedom, the sum of squares for each contrast adds up to the sum of squares for treatments. Thus, if the null hypothesis for treatments is rejected, the sum of squares for treatments can be partitioned into the sums of squares for each individual contrast and the reason for the rejection of H_0 can be found.

It is important that the $k - 1$ contrasts be identified *before* the experiment is run. This is because contrasts suggested by the data, after the experiment is complete, can lead to misleading conclusions and an increase in type II errors. (Reread the second paragraph of this chapter.) The experimenter who wants to look at comparisons suggested by the data should use the techniques of the next section on multiple comparisons.

While the mutual orthogonality of the contrasts is desirable, it is not essential and the experimenter who finds some nonorthogonal contrasts interesting should proceed as long as the contrasts were suggested before the experiment was run. However, it should be noted that the sum of squares of the contrasts will not add up to the treatment sum of squares and the F-ratios may not be independent.

Example 10.5 In example 10.1 suppose that chemical 1 is a standard against which we wish to compare the others. Since there are three treatments, we can choose two orthogonal contrasts to partition the treatment sum of squares. Suppose we wish to test

$$H_0\colon 2\mu_1 - \mu_2 - \mu_3 = 0$$

and

$$H_0\colon \quad \mu_2 - \mu_3 = 0$$

Notice that if the second hypothesis had been $H_0\colon \mu_1 - \mu_2 = 0$, the contrasts would not have been orthogonal. The two contrasts are

$$C_1 = 2\overline{X}_1 - \overline{X}_2 - \overline{X}_3, \quad C_2 = \overline{X}_2 - \overline{X}_3$$

for which

$$SS(C_1) = \frac{4(-1.5)^2}{6} = 1.5, \qquad SS(C_2) = \frac{4(4.5)^2}{2} = 40.5$$

We see that SSTR = $SS(C_1) + SS(C_2)$ as stated. The F-ratios are

$$F_1 = \frac{1.5}{3.22} = 0.47, \qquad F_2 = \frac{40.5}{3.22} = 12.78$$

and $F_{.05}$ (with 1 and 9 degrees of freedom) is 5.12. We reject the second hypothesis and conclude that μ_2 and μ_3 are different but that we are not able to detect a difference between μ_1 and the average of μ_2 and μ_3. The results can be combined in the ANOVA table as follows:

Source of variation	Sum of squares	Degrees of freedom	Mean square	F-ratio
Between	42.0	2	21.0	
μ_1 vs (μ_2,μ_3)	1.5	1	1.5	0.47
μ_2 vs μ_3	41.5	1	40.5	12.78
Within	29.0	9	3.22	
Total	71.0	11		

10.1.7 Multiple Comparisons

Contrasts suggested after the data have been collected instead of before, are often referred to as "data-snooping" devices or as *multiple comparison* methods. There are many such multiple comparison methods of which we shall mention just two. We start by considering *Tukey's test.*

Suppose that there are k samples each having n observations. The sample means are arranged in ascending order (ranked) $\overline{X}_{(1)}$

$< \overline{X}_{(2)} < \ldots < \overline{X}_{(k)}$, where $\overline{X}_{(i)}$ is the *i*th smallest. A constant, H, called the significant difference, is calculated as follows:

$$H = q(\alpha, k, k(n - 1))\sqrt{MSE/n}$$

where $q(\alpha, k, k(n - 1))$ is a value tabulated in Appendix Table 11, that depends on α, k, and $k(n - 1)$. If the absolute value of the difference of any pair of means is greater than H, that pair of means is declared to be significantly different.

Tukey's test keeps the probability of a type I error small and, hence, is often considered too conservative by many researchers, who would rather pick up too many differences than too few.

Example 10.6 Suppose that a CRD was run with 7 treatments each having 6 observations. The F-ratio was significant at the 0.05 level of significance and MSE = 79.64. The ranked means are as follows:

Ranked mean	1	2	3	4	5	6	7
	49.6	58.1	61.0	61.5	67.6	71.2	71.3

Use Tukey's test to separate the means. From Appendix Table 11,

$$q(.05, 7, 35) \cong 4.43$$

so that H = 16.14. The results are given below; those means underlined with the same line are considered to be not significantly different.

Mean	1	2	3	4	5	6	7
	49.6	58.1	61.0	61.5	67.6	71.2	71.3

Another method of separating means is called *Duncan's multiple range test.* Again, the k treatments each having n observations are arranged in ascending order. We say that $\overline{X}_{(1)}$ and $\overline{X}_{(k)}$ are k steps apart; $\overline{X}_{(2)}$ and $\overline{X}_{(k)}$ are $k - 1$ steps apart, etc. Duncan's test uses a protection level α_p instead of a significance level α. In general, α will be larger than α_p. Consider the comparison of two means p steps apart. Duncan's method compares the positive difference of two means against

$$W_p = r(\alpha_p, p, k(n - 1)) \sqrt{MSE/n}$$

where $r(\alpha, p, k(n - 1))$ is tabulated in Appendix Table 12. Its value depends on α_p and $k(n - 1)$, but also on p, the number of means being compared. If the difference between two means is larger than W_p, the two means are declared to be significantly different. With a protection level of α_p, the level of significance (probability of a type I error) is considerably higher than α_p. This makes Duncan's test more capable of detecting differences. For this reason it has been readily accepted among researchers, especially in agriculture, where we are more interested in finding differences than in not finding them. It is usual to require a significant F-test in the analysis of variance before applying Duncan's test.

To perform the test, the means are ordered from smallest to largest and each pair is compared to W_p starting with the two extremes. If two means are declared to be not significantly different, they are underlined with the same line and no further testing need be made of intermediate means.

Example 10.7 Analyze the results of example 10.6 using Duncan's multiple range test.

We shall require W_p for p = 2, 3, 4, 5, 6, and 7. From Appendix Table 12 the values of r(.05, p, 35) are (approximately):

p	r(.05, p, 35)	$W_p = r\sqrt{MSE/n}$
2	2.87	10.46
3	3.02	11.00
4	3.11	11.37
5	3.19	11.62
6	3.24	11.80
7	3.28	11.95

The difference $\overline{X}_{(7)} - \overline{X}_{(1)}$ is 21.7 which is greater than 11.95 and declared significant. Next $\overline{X}_{(6)} - \overline{X}_{(1)} = 21.6$ and $\overline{X}_{(7)} - \overline{X}_{(2)} = 13.2$ are declared significant, being greater than 11.80. This process is continued with the following results:

Mean	1	2	3	4	5	6	7
	49.6	58.1	61.0	61.5	67.6	71.2	71.3

10.1.8 Unequal Sample Sizes

Equal sample sizes are important in any experimental design, and the experimenter should try, insofar as possible, to achieve this. We refer to such designs as *balanced.* However, since balanced designs are not always possible, we must consider what to do with a CRD with unequal sample sizes (called an *unbalanced* design).

Suppose that there are n_i observations in the *i*th sample and that the total number of observations is $n_1 + n_2 + \ldots + n_k$, denoted N. The sum of squares are

$$SST = \sum_{i=1}^{k} \sum_{j=1}^{n_i} (X_{ij} - \overline{X})^2 = \Sigma\Sigma X_{ij}^2 - T^2/N$$

$$SSTR = \sum_{i=1}^{k} n_i(\overline{X}_i - \overline{X})^2 = \sum_{i=1}^{k} \left(\frac{T_i^2}{n_i}\right) - T^2/N$$

$$SSE = \sum_{i=1}^{k} \sum_{j=1}^{n_i} (X_{ij} - \overline{X}_i)^2 = \Sigma\Sigma X_{ij}^2 - \sum_{i=1}^{k} \left(\frac{T_i^2}{n_i}\right)$$

The degrees of freedom for treatments, error, and total are $k - 1$, $N - k$, and $N - 1$, respectively. The mean squares and F-ratio are calculated as previously. The ANOVA table 10.4 is shown below.

Table 10.4 *ANOVA table for an unbalanced CRD.*

Source of variation	Sum of squares	Degrees of freedom	Mean square	F-ratio
Between	SSTR	$k - 1$	MSTR	MSTR/MSE
Within	SSE	$N - k$	MSE	
Total	SST	$N - 1$		

To test H_0: $\mu_1 = \mu_2 = \ldots = \mu_k$ at significance level α, we reject H_0 if the F-ratio is greater than F_α with $f_1 = k - 1$ degrees of freedom and $f_2 = N - k$ degrees of freedom. Since the variance of $\overline{X}_i$ is now σ^2/n_i, a $100(1 - \alpha)\%$ confidence interval for μ_i is

$$\left(\overline{X}_i - t_{\alpha/2}\sqrt{\frac{MSE}{n_i}},\ \overline{X}_i + t_{\alpha/2}\sqrt{\frac{MSE}{n_i}}\right)$$

and a $100(1 - \alpha)\%$ confidence interval for $\mu_i - \mu_j$ is

$$\left(\overline{X}_i - \overline{X}_j - t_{\alpha/2}\sqrt{MSE\left(\frac{1}{n_i} + \frac{1}{n_j}\right)},\ \overline{X}_i - \overline{X}_j + t_{\alpha/2}\sqrt{MSE\left(\frac{1}{n_i} + \frac{1}{n_j}\right)}\right)$$

In terms of the mathematical model, MSE is an unbiased estimator of σ^2, while MSTR is an unbiased estimator of

$$\sigma^2 + \frac{1}{(k - 1)^?}\left(N - \frac{\sum_{i=1}^{k} n_i^2}{N}\right)\Sigma\, \tau_i^2$$

The critical value of H for comparing the differences of any pair of means is

$$H = \frac{q(\alpha, k, N - k)\sqrt{MSE}}{\min(\sqrt{n_i})}$$

where $\min(\sqrt{n_i})$ is the smallest square root of the n_i's. To use Duncan's test for the case where the *i*th sample has n_i observations, it has been suggested that a good approximation results from replacing MSE/n by

$$\frac{MSE}{2}\left(\frac{n_i + n_j}{n_i n_j}\right)$$

Example 10.8 The following figures are the results of an experiment in which the growth rate in mm/month for a particular vegetable is compared under four different growing conditions. The vegetables were randomly assigned to growing condition, resulting in a completely randomized design. Analyze the results and test the hypothesis that growth rates are the same for all growing conditions (with $\alpha = 0.05$). Set a 90% confidence interval for $\mu_2 - \mu_1$, and separate the means using Tukey's and Duncan's methods.

	Condition 1	Condition 2	Condition 3	Condition 4
	63	70	68	63
	54	74	46	85
	47	51	49	80
	32		53	95
				82
Sums	196	195	216	405
Means	49	65	54	81
n_i	4	3	4	5

$T = 1012,\ N = 16$

$$SST = 63^2 + \ldots + 82^2 - (1012)^2/16 = 4379$$

$$SSTR = \frac{196^2}{4} + \ldots + \frac{405^2}{5} - (1012)^2/16 = 2639$$

$$SSE = 63^2 + \ldots + 82^2 - \frac{196^2}{4} + \ldots + \frac{405^2}{5} = 1740$$

Source of variation	Sum of squares	Degrees of freedom	Mean square	F-Ratio
Between conditions	2639	3	879.67	6.07
Within samples	1740	12	145.00	
Total	4379	15		

Since $F_\alpha = F_{.05} = 3.49$, the null hypothesis is rejected. The conclusion is that the growth rates are affected by differing conditions. A 90% confidence interval for $\mu_2 - \mu_1$ is

$$(16 - (1.78)\sqrt{84.58},\ 16 + (1.78)\sqrt{84.58})$$

or $(-0.40, 32.40)$.

To use Tukey's method

$$H = \frac{q(.05, 4, 12)\sqrt{MSE}}{\min(\sqrt{n_i})} = \frac{(4.20)(12.04)}{1.732} = 29.20$$

with the following results:

Condition	1	3	2	4
Mean	49	54	65	81

Use Duncan's test as follows

p	r(.05, p, 12)
2	3.08
3	3.23
4	3.31

To compare means 1 and 3, $n_1 = 4$, $n_3 = 4$, and $W_p = 18.54$.
To compare means 1 and 4, $n_1 = 4$, $n_4 = 5$, and $W_p = 18.91$, etc.

Condition	1	3	2	4
Mean	49	54	65	81

10.1.9 The Random Effect Model

The difference between the fixed and random effects model was discussed in section 10.1.3. In fact, if we are dealing with a random effects model, the sum of squares, degrees of freedom, and mean squares are the same as in sections 10.1 and 10.1.8. However, in the model

$$X_{ij} = \mu + \tau_i + \varepsilon_{ij}$$

the τ_i's are assumed to be random variables from a normal population with mean zero and variance σ_τ^2, while the ε_{ij}'s are assumed to be (independent) random variables from a normal distribution with mean 0 and variance σ^2. We interpret σ^2 as the variance within each sample and σ_τ^2 as the variance of the population effects τ_i. It can be shown that for the random effects model, MSE is an unbiased estimator of σ^2, while MSTR is an unbiased estimator of

$$n_0 = \frac{1}{k-1}\left(N - \frac{\sum_{i=1}^{k} n_i^2}{N}\right)$$

which reduces to n for equal sample sizes. An unbiased estimator S^2 of σ^2 is thus MSE, while an unbiased estimator S_τ^2 of σ_τ^2 is (MSTR − MSE)/n_0.

Example 10.9 For simplicity, suppose that the four conditions of example 10.6 are a random sample from a large population of possible conditions, so that a random effects model is appropriate. With MSE = 145, MSTR = 879.67, an unbiased estimate of σ^2 is

$$S^2 = 145.00$$

The value of n_0 is

$$n_0 = \frac{1}{3}\left(16 - \frac{[4^2 + 3^2 + 4^2 + 5^2]}{16}\right) = 3.958$$

so that an estimate of σ_τ^2 is $S_\tau^2 = 185.62$.

10.1.10 Testing the Homogeneity of Sample Variances

One of the basic assumptions of the analysis of variance technique for both fixed and random effects models is that the variances within each population are the same. If the populations are labeled 1, 2, . . . , k with means $\mu_1, \mu_2, \ldots, \mu_k$ and variances $\sigma_1^2, \sigma_2^2, \ldots,$ σ_k^2, this amounts to assuming that $\sigma_1^2 = \sigma_2^2 = \sigma_k^2$ (σ^2). The denominator MSE of the F-ratio estimates σ^2 by pooling together the variance estimators from each of the samples. If the variances σ_1^2, σ_2^2, . . . , σ_k^2 are not equal, then the F-ratio does not have an F-distribution, and tests involving the F-distribution will be invalid. If the variances are not too different, the F-ratio will still have approximately a tabulated F-distribution.

If we wish to test that the variances are equal, we can use a number of tests, two of which we will discuss here. The simplest is the *Hartley test* and requires that the k samples have the same number of observations n. If the k sample variances are S_1^2, S_2^2, . . . , S_k^2 where

$$S_i^2 = \frac{\sum_{j=1}^{n} (X_{ij} - \overline{X}_i)^2}{n - 1}, \quad \overline{X}_i = \frac{1}{n}\sum_{j=1}^{n} X_{ij}$$

After being arranged from smallest to largest, S^2_{min} is the smallest and S^2_{max} is the largest, and the Hartley's test statistic for testing H_0: $\sigma_1^2 = \sigma_2^2 = \ldots = \sigma_k^2$ is

$$H = S^2_{max} / S^2_{min}$$

The distribution of H, which depends on the values of k and $\upsilon = n - 1$, has been tabulated in Appendix Table 13. In particular, if H_α is the point on the distribution of H for which

$$P[H > H_\alpha] = \alpha$$

we reject the null hypothesis at α level of significance if $H > H_\alpha$. If the sample sizes are not equal but not too different, we may calculate H as before. However, the value of H_α is read from the table for values k and the largest of the sample sizes. Such a procedure leads to a slight increase in the type I error probability but will generally be adequate.

Example 10.10 Use Hartley's test on the data of example 10.1 to see if the variances are homogeneous. Use a significance level of 0.05.

$$S_1^2 = 2, S_2^2 = 2, S_3^2 = 5.67, \upsilon = 3, k = 3$$

so that H = 2.83. From Appendix Table 13 $H_{.05} = 27.80$. Since H is not greater than $H_{.05}$, we do not reject H_0. There is no evidence to suggest that the variances are unequal.

The best known test for homogeneity of variances is *Bartlett's* test. If the *i*th sample contains n_i observations, and S_i^2 is the sample variance, and if

$$C = 1 + \frac{1}{3(k-1)}\left\{\sum_{i=1}^{k}\left(\frac{1}{n_i - 1}\right) - \frac{1}{\Sigma(n_i - a)}\right\}$$

then Bartlett's test statistic is

$$B = \frac{1}{C}\left\{\sum_{i=1}^{k}(n_i - 1)\log \text{MSE} - \sum_{i=1}^{k}(n_i - 1)\log S_i^2\right\}$$

where we use the natural logarithm. The distribution of B has approximately a χ^2-distribution with $k - 1$ degrees of freedom. To test H_0: $\sigma_1^2 = \sigma_2^2 = \ldots = \sigma_k^2$ at the α level of significance, we reject H_0 if $B > \chi_\alpha^2$.

Although Bartlett's test does not need equal sample sizes and is therefore more flexible than Hartley's test, its use is not to be recommended, since the distribution of B is very sensitive to the assumption that the populations from which the samples are drawn are normal. If this is not so, invalid conclusions may be reached.

Example 10.11 Use Bartlett's test on the data of example 10.1 to see if the variances are homogeneous. Use $\alpha = 0.05$.

$$C = 1 + \frac{1}{6}\left\{\frac{1}{4} + \frac{1}{4} + \frac{1}{4} - \frac{1}{9}\right\} = 1.106$$

$$\Sigma\,(n_i - 1)\log \text{MSE} = 9(1.169) = 10.524$$

$$\Sigma\,(n_i - 1)\log S_i^2 = 3(0.693 + 0.093 + 1.735) = 9.363$$

$$B = \frac{1}{1.106}(1.161) = 1.05$$

Since $\chi_{.05}^2$ with 2 degrees of freedom is 5.99, the null hypothesis cannot be rejected.

10.1.11 Data Transformation

To perform the analysis of variance validly, we assume that the data are normally distributed with equal sample variances. A test for normality was discussed in chapter 7 and tests for homogeneity of variance were discussed in section 10.1.10. Whether these assumptions are satisfied or not depends on the scale of measurements used. Where one or the other assumption is violated, it may be possible to choose an alternate scale wherein the assumptions will be satisfied. A change of scale is called a *transformation* of the data. Some of the more common transformations concern logarithms, square roots, reciprocals, and arcsines.

Recent research shows that the distribution of the F-ratios is not overly sensitive to variance inequalities pooled into the error variance or to moderate departures from normality. Thus, transformations that have the objective of homogenizing the variances or attaining normality are less important than was earlier thought. Nevertheless, it is felt that if the data need transforming, the following guidelines can be useful:

1. If the sample variance for each treatment is proportional to the mean of that treatment (i.e., the larger the treatment mean, the larger the variance), the square root transformation (from x_{ij} to $\sqrt{x_{ij}}$) will help stabilize the variance.
2. If the sample variances are proportional to the square of the sample means, the logarithmic transformation (from x_{ij} to $\log x_{ij}$ or $\log(x_{ij} + 1)$) will help stabilize the variance.
3. If the observations are proportions or percentages, if we transform from x_{ij} to

$$2 \text{ arcsine } (\sqrt{x_{ij}})$$

this will help stabilize the variance.

In other cases, finding an appropriate transformation is a trial-and-error procedure, best aided by high-speed computers. One unified approach (called the method of Box and Cox) is to consider transforming from x_{ij} to y_{ij} where

$$y_{ij} = \begin{cases} \dfrac{(x_{ij} + A)^B - 1}{B} & \text{for } B \neq 0 \\ \log(x_{ij} + A) & \text{for } B = 0 \end{cases}$$

where A and B are chosen such that

$$L = -\frac{n}{2}\log(s^2) + (B - 1)\sum_{i=1}^{n}\log(x_{ij} + A)$$

is maximized. We search for the maximizing values of A and B by trial-and-error. If A′ and B′ are the two values thus found, the transformed data are

$$y_{ij} = \frac{(x_{ij} + A')^{B'} - 1}{B'} \qquad \text{if } B' \neq 0$$

or

$$y_{ij} = \log(x_{ij} + A') \qquad \text{if } B' = 0$$

This approach optimizes data normality and variance homogeneity at the same time.

10.2 Randomized Complete Block Designs

We saw, in chapter 7, that if observations were paired (such as a before and after measurement on the same subject or object), a paired t-test was used to test H_0: $\mu_1 = \mu_2$. Each pair of observations was treated as a group or a *block.* Observations within a block were positively correlated and more nearly alike than observations from two separate blocks. This idea of blocking may be extended to the comparison of k population means ($k > 2$) by using the *randomized complete block design (RCBD).* This design is used when observations are naturally arranged into groups and when the observations within a group are more nearly alike or more homogeneous than observations taken from different groups. The RCBD is extremely useful in agricultural and food experiments if some observations are obtained under one set of conditions and others are obtained under other conditions. For instance, plants grown indoors during one season may form one block, and plants grown outside during another season may form another block. Other sources of changing conditions are weather and soil variability. We may recall from the CRD that the denominator of the F-ratio is the mean square error which estimates the within-sample variance. If this variance is inflated because of uncontrollable conditions (weather, soil type, or elapsed time), the F-ratio will be smaller, the null hypothesis will be less likely to be rejected, and the probability of a type II error will be greater.

The experimenter needs to control as much extraneous variability as possible to minimize the MSE. He can also statistically control the variability by using the RCBD. Successful use of the RCBD depends upon the experimenter's ability to identify the sources of extraneous variability. For instance, a common source of variability in field experiments is soil gradient. In such a case the experiment could be repeated for each of the different soil types. The blocks would then correspond to soil type. Suppose that there are k treatments to be compared. Each of the n blocks would be split into k plots, and each treatment would be applied exactly once per block. The treatments are applied randomly within a block and each treatment occurs exactly once per block.

For n blocks, each of size k, there are nk total plots or observations. Once the data have been collected, they may be displayed as shown here:

		TREATMENT				
		1	2	. . .	k	Totals
BLOCK	1	X_{11}	X_{12}	. . .	X_{1k}	B_1
	2	X_{21}	X_{22}	. . .	X_{2k}	B_2
		.				
		.				
		.				
	n	X_{n1}	X_{n2}	. . .	X_{nk}	B_n
	Totals	T_1	T_2	. . .	T_k	T

where X_{ij} = observation in block i, treatment j

B_i = sum of observations in block i

T_j = sum of observations in treatment j

T = grand total

$\overline{B}_i$ = mean of block i

$\overline{T}_j$ = mean of treatment j

$\overline{X}$ = grand mean

To partition the total sum of squares into its components, note that the total sum of squares is the sum of the squares of $X_{ij} - \overline{X}$. But,

$$X_{ij} - \overline{X} = (X_{ij} - \overline{B}_i - \overline{T}_j + \overline{X}) + (\overline{B}_i - \overline{X}) + (\overline{T}_j - \overline{X})$$

which, when squared and summed, gives

$$\sum_{i=1}^{n}\sum_{j=1}^{k} (X_{ij} - \overline{X})^2 = \sum_{i}\sum_{j} (X_{ij} - \overline{B}_i - \overline{T}_j + \overline{X})^2$$

$$+ \sum_{i=1}^{n} k(\overline{B}_i - \overline{X}) + \sum_{j=1}^{k} n(\overline{T}_j - \overline{X})^2$$

since all crossproduct terms can be shown after some algebraic manipulation to be equal to 0.

$\sum_{i}\sum_{j} (X_{ij} - \overline{X})^2$ is called the total sum of squares (SST)

$\sum_{i} k(\overline{B}_i - \overline{X})^2$ is called the block sum of squares (SSB)

$\sum_{j} n(\overline{T}_j - \overline{X})^2$ is called the treatment sum of squares (SSTR)

$\sum_{i}\sum_{j} (X_{ij} - \overline{B}_i - \overline{T}_j + \overline{X})^2$ is called the error sum of squares (SSE).

Notice that SSB measures the variation of block means about the grand mean and the SSTR measures the variation of treatment means about the grand mean. The remainder of the variation called the SSE is assumed to be due to experimental variation. Computationally more convenient formulas are as follows:

$$\text{SST} = \sum_{i=1}^{n}\sum_{j=1}^{k} X_{ij}^2 - T^2/nk$$

$$\text{SSB} = \frac{1}{k}\sum_{i=1}^{n} B_i^2 - T^2/kn$$

$$SSTR = \frac{1}{n}\sum_{j=1}^{k} T_j^2 - T^2/kn$$

$$SSE = SST - SSB - SSTR$$

The degrees of freedom for SST, SSB, SSTR, and SSE are nk − 1, n − 1, k − 1, and (n − 1)(k − 1), respectively. The ANOVA table may be written as follows:

Source of variation	Sum of squares	Degrees of freedom	Mean square
Blocks	SSB	n − 1	MSB = SSB/n − 1
Treatments	SSTR	k − 1	MSTR = SSTR/k − 1
Error	SSE	(n − 1)(k − 1)	MSE = SSE/(n − 1)(k − 1)
Total	SST	nk − 1	

Example 10.12 Three different oils A, B, and C were to be compared for their ability to prevent baked cakes from sticking to the baking dish. To ensure uniformity, a single baking mixture was created and 3 similar dishes were coated with oils A, B, and C, respectively. The response was measured on a 0–10 scale by a group of experts. They wanted to repeat the process 4 times for each oil to give a total of 12 observations. Unfortunately, the ovens were not large enough to take all 12 dishes and it was decided to use 4 ovens, each one being treated as a block. In each oven were placed 3 dishes with oils A, B, and C. The following results were obtained. Analyze them using a RCBD.

	TREATMENT (Oil) A	B	C	Totals
BLOCKS (Ovens) 1	6.2	5.7	2.4	14.3
2	4.2	4.0	1.8	10.0
3	8.6	9.3	7.7	25.6
4	7.6	8.5	5.1	21.2
Totals	26.6	27.5	17.0	71.1

The sums of squares are as follows.

$$\begin{aligned} SST &= 489.33 - 421.26 = 68.07 \\ SSB &= 469.76 - 421.26 = 48.50 \\ SSTR &= 438.20 - 421.26 = 16.94 \\ SSE &= SST - SSB - SSTR = 2.63 \end{aligned}$$

and the ANOVA table is

Source	Sum of squares	Degrees of freedom	Mean square
Blocks	48.50	3	16.17
Treatments	16.94	2	8.47
Error	2.63	6	0.44
Total	68.07	11	

In section 10.1.3, the mathematical model for a CRD was given in which the observations were composed of an overall mean μ, a deviation from the overall mean τ_i due to treatment i, and a normal random variable ε_{ij}. For the RCBD we assume that there is an overall mean μ, a deviation from the mean β_i due to block i, a deviation from the mean τ_j due to treatment j, and a normal random variable ε_{ij}. This is illustrated with the following data (where the random error is ignored):

		TREATMENT 1	2	3	Mean
BLOCK	1	108	103	104	105
	2	104	99	100	101
	3	99	94	95	96
	4	101	96	97	98
Mean		103	98	99	100

where $\mu = 100$, $\tau_1 = 3$, $\tau_2 = -2$, $\tau_3 = -1$, $\beta_1 = 5$, $\beta_2 = 1$, $\beta_3 = -4$, $\beta_4 = -2$. Since τ_j and β_i are deviations from the overall mean μ, they sum to zero. The mathematical model for the RCBD is thus

$$X_{ij} = \mu + \beta_i + \tau_j + \varepsilon_{ij} \text{ for } i = 1, 2, \ldots, n \text{ and } j = 1, 2, \ldots, k$$

where X_{ij} is the observation for block i, treatment j

μ is the overall mean

β_i is the deviation from μ due to block i

τ_j is the deviation from μ due to treatment j

ε_{ij} is the normal random variable in block i, treatment j, having variance σ^2

If the β_i's and τ_j's are assumed to be fixed effects, MSE is an unbiased estimator of σ^2, MSB is an unbiased estimator of

$$\sigma^2 + \frac{k}{n-1}\sum_{i=1}^{n} \beta_i^2$$

and MSTR is an unbiased estimator of

$$\sigma^2 + \frac{n}{k-1}\sum_{j=1}^{k} \tau_j^2$$

In some biological situations, blocks are regarded as samples from a larger population, in which case block is a random effect for which $\beta_i \sim N(0, \sigma_\beta^2)$ and MSB is an unbiased estimator of $\sigma^2 + k\,\sigma_\beta^2$. Neither the partitioning of the sums of squares nor subsequent tests of hypotheses are affected by this assumption.

The most important hypothesis to be tested with the RCBD is $H_0: \tau_1 = \tau_2 = \ldots = \tau_k$, which, if true, implies that there are no effects of treatments. This is tested with the F-ratio

$$F = MSTR/MSE$$

Remembering that MSTR estimates the variance between treatment means and that MSE estimates the experimental (or error) variance, F is the ratio of two variances and has an F-distribution with $f_1 = k - 1$ degrees of freedom and $f_2 = (k - 1)(n - 1)$ degrees of freedom. If the null hypothesis of equal treatment means is true, the variances between means and error will be the same and the ratio MSTR/MSE will be close to 1. If, on the other hand, the null hypothesis is not true and at least one treatment mean (and hence τ_j) is different from the others, the ratio MSTR/MSE will be greater than 1. Hence, large values of F indicate that the null hypothesis is false. In fact, we reject the null hypothesis at the α level of significance if $F > F_\alpha$, where F_α is read in Appendix Table 10 with f_1 and f_2 degrees of freedom. A $100(1 - \alpha)\%$ confidence interval for treatment μ_i is

$$\left(\overline{T}_i - t_{\alpha/2}\sqrt{\frac{MSE}{n}},\ \overline{T}_i + t_{\alpha/2}\sqrt{\frac{MSE}{n}}\right)$$

and a $100(1 - \alpha)\%$ confidence interval for the difference of two treatment means $\mu_i - \mu_j$ is

$$\left(\overline{T}_i - \overline{T}_j - t_{\alpha/2}\sqrt{\frac{2MSE}{n}},\ \overline{T}_i - \overline{T}_j + t_{\alpha/2}\sqrt{\frac{2MSE}{n}}\right)$$

where the degrees of freedom for $t_{\alpha/2}$ are the same as for error.

Example 10.13 Continuing with example 10.12, test the hypothesis of equal effects (with $\alpha = 0.05$) of the three oils. Then calculate a 95% confidence interval for the effect of oil A, and a 95% confidence interval for the difference in effect of oils B and C.

The F-ratio is MSTR/MSE = 19.25. The tabulated value of $F_{.05}$ with 2 and 6 degrees of freedom is 5.14. We conclude that the oils do have differing effects in their abilities to prevent sticking. A 95% confidence interval for the scores for oil A is

$$\left(26.6 - (2.45)\sqrt{\frac{0.44}{4}},\ 26.6 + (2.45)\sqrt{\frac{0.44}{4}}\right)$$

or (25.79, 27.41)
A 95% confidence interval for the difference of oils B and C is

$$\left(10.5 - (2.45)\sqrt{\frac{0.88}{4}},\ 10.5 + (2.45)\sqrt{\frac{0.88}{4}}\right)$$

or (9.35, 11.65).

We can analyze the treatment effects further by using orthogonal contrasts (as in section 10.1.6) or by using multiple comparison (as in section 10.1.7).

Example 10.14 Using the data of example 10.12, suppose that before the experiment started we stated our wish to compare A against B and C and also to compare B with C. We wish to test

$$H_0\text{: } 2\tau_A - \tau_B - \tau_C = 0$$
$$H_0\text{: } \tau_B - \tau_C = 0$$

so that $C_1 = 2\overline{T}_A - \overline{T}_B - \overline{T}_C$, $C_2 = \overline{T}_B - \overline{T}_C$

$$SS(C_1) = \frac{4(2.175)^2}{6} = 3.15,\ SS(C_2) = \frac{4(2.625)^2}{2} = 13.77$$

$$F_1 = \frac{3.15}{0.44} = 7.20,\ F_2 = \frac{13.77}{0.44} = 31.30$$

and $F_{.05}$ (with 1 and 6 degrees of freedom) is 5.99. Since both F_1 and F_2 are greater than $F_{.05}$, we conclude that both contrasts are significantly different, so that the mean response for oil A is different from the means of oils B and C and, further, B and C are themselves different.

Suppose in this same example we wish to use Tukey's test to separate the means.

$$H = q(0.05, 3, 9)\sqrt{MSE/4} - (3.95)\sqrt{0.44/4} = 1.31$$

The treatments are ranked below. Those underlined with the same line are regarded as being not significantly different.

Oil	B	A	C
Mean	<u>6.875</u>	<u>6.65</u>	4.25

10.2.1 Missing Observations

The RCBD is balanced in the sense that there is exactly one observation per cell. Occasionally we have to deal with an RCBD with one observation missing. This might be caused by the loss of an animal or plant, a blunder on the part of a researcher, or any one of a number of other possible reasons. Suppose that the missing observation is X_{ij} (i.e., in block i and treatment j). It can be estimated by

$$X_{ij} = \frac{nB_i + kT_j - T}{(k - 1)(n - 1)}$$

where B_i = sum of observations in block i

T_j = sum of observations in treatment j

T = sum of all observations.

The treatment sum of squares and the total sum of squares are reduced by

$$\frac{[B_i - (k - 1)X_{ij}]^2}{k(k - 1)}$$

The total degrees of freedom and the error degrees of freedom are reduced by 1. Discussion of the procedure to be followed when two observations are missing can be found in Steel and Torrie, 1960.

Example 10.15 Reconsider example 10.12 and suppose that the first observation (6.2) is missing due to an experimental blunder. Reanalyze the experiment. The data available are

		TREATMENT (Oil)			
		A	B	C	Totals
BLOCK	1		5.7	2.4	8.1
	2	4.2	4.0	1.8	10.0
	3	8.6	9.3	7.7	25.6
	4	7.6	8.5	5.1	21.2
Totals		20.4	27.5	17.0	64.9

The estimate of X_{11}, the missing point is

$$X_{11} = \frac{(4)(8.1) + (3)(20.4) - 64.9}{(2)(3)} = 4.78$$

The treatment sum of squares is reduced by

$$\frac{[8.1 - (2)(4.78)]^2}{6} = 0.36$$

The ANOVA table is

Source	Sum of squares	Degrees of freedom	Mean square	Ratio
Blocks	48.50	3	16.17	
Treatments	16.58	2	8.29	15.9
Error	2.63	5	0.52	
Total	67.71	10		

10.3 Relative Efficiency of CRD and RCBD

We use a RCBD to remove variability due to block effects from the error variance. If block effects are present and a CRD is used, the error variance (used to estimate σ^2) will be inflated owing to the presence of the block effect. We shall consider an experiment designed as a RCBD but analyzed also as a CRD and see what effect the changes in the error variance and degrees of freedom have. The sums of squares, degrees of freedom, and mean squares are shown:

Completely Randomized Design

Source	Sum of squares	Degrees of freedom	Mean square
Treatments	$\Sigma n(\overline{X}_i - \overline{X})^2 = SSTR$	$k - 1$	$SSTR/k - 1$
Error	$\Sigma\Sigma(X_{ij} - \overline{X}_i)^2 = SSE$	$k(n - 1)$	$SSE/k(n - 1)$
Total	$\Sigma\Sigma(X_{ij} - \overline{X})^2 = SST$	$kn - 1$	

Randomized Complete Block Design

Source	Sum of squares	Degrees of freedom	Mean squares
Treatments	$\Sigma n(\overline{T}_j - \overline{X})^2 = SSTR$	$k - 1$	$SSTR/k - 1$
Blocks	$\Sigma k(\overline{B} - \overline{X})^2 = SSB$	$n - 1$	$SSB/n - 1$
Error	$\Sigma\Sigma(X_{ij} - \overline{B}_i - \overline{T}_j + \overline{X})^2 = SSE$	$(k - 1)(n - 1)$	$SSE/(k - 1)(n - 1)$
Total	$\Sigma\Sigma(X_{ij} - \overline{X})^2 = SST$	$kn - 1$	

Since $\overline{X}_i$ in the CRD is the same as $\overline{T}_j$ in the RCBD, we see immediately that the SST and the SSTR, along with their corresponding degrees of freedom, are the same in both designs. Thus, the SSE of the CRD is the sum of the error and block sum of squares for the RCBD (as are the degrees of freedom). Suppose that for the CRD the error mean square (which we shall call M_1) is an estimator of σ_1^2. From section 10.1.4, MSTR is an estimator of

$$\sigma_1^2 + \frac{n}{k-1}\Sigma\tau_i^2$$

so that the total sum of squares is an estimator of

$$(k-1)\left[\sigma_1^2 + \frac{n}{k-1}\Sigma\tau_i^2\right] + k(n-1)\sigma_1^2$$

If for the RCBD, the error mean square (which we shall call M_2) is an estimator of σ_2^2, from section 10.2, MSB (which we shall call M_3) is an estimator of

$$\sigma_2^2 + \frac{k}{n-1}\Sigma\beta_i^2$$

and MSTR is an estimator of

$$\sigma_2^2 + \frac{n}{k-1}\Sigma\tau_i^2$$

so that the total sum of squares is an estimator of

$$(n-1)\left[\sigma_2^2 + \frac{k}{n-1}\Sigma\beta_i^2\right] + (k-1)\left[\sigma_2^2 + \frac{n}{k-1}\Sigma\tau_i^2\right] + (n-1)(k-1)\,\sigma_2^2$$

Since the total sums of squares for the CRD and the RCBD are equal, we can equate the two expressions for the total sums of squares and obtain

$$\sigma_1^2 = \sigma_2^2 + \frac{k}{nk-1}\Sigma\beta_i^2$$

Since M_2 and M_3 are estimators of σ_2^2 and

$$\sigma_2^2 + \frac{k}{n - 1} \Sigma\beta_i^2$$

respectively, it can be shown that $(M_3 - M_2)(n - 1)/k$ is an estimator of $\Sigma\beta_i^2$. Putting these into

$$\sigma_1^2 = \sigma_2^2 + \frac{k}{kn - 1} \Sigma\beta_i^2$$

yields $M_1 = \dfrac{n(k - 1)\, M_2 + (n - 1)\, M_3}{nk - 1}$

The ratio M_1/M_2 can be used as a test for the efficiency of the RCBD relative to the CRD. Whenever the ratio M_1/M_2 is greater than 1, the reduction in the estimate of σ^2 more than compensates for the loss in degrees of freedom in the RCBD. Hence, ratios larger than 1 justify the use of the RCBD.

An improved measure of efficiency of the two designs which takes into account the different error degrees of freedom (but which we will not derive) is

$$\text{Efficiency} = M_1 \frac{(f_1 + 3)}{(f_1 + 1)} \Big/ M_2 \frac{(f_2 + 3)}{(f_2 + 1)}$$

where f_1 and f_2 are the error degrees of freedom for the CRD and the RCBD, respectively.

Example 10.16 For the data of example 10.12 $M_2 = 0.44$, $M_3 = 16.17$, $f_1 = 9$, $f_2 = 6$, so that

$$M_1 = \frac{(4)(2)(0.44) + (3)(16.17)}{11} = 4.73$$

$$M_1/M_2 = 10.75 \quad \text{and} \quad M_1 \frac{f_1 + 3}{f_1 + 1} \Big/ M_2 \frac{f_2 + 3}{f_2 + 1} = 10.03$$

We conclude that the use of the RCBD was justified.

Exercises

1. Four varieties of wheat are grown in twelve plots of land (three plots per variety) distributed randomly and the yield per acre measured. The following data were obtained:

Variety 1	Variety 2	Variety 3	Variety 4
53	57	43	61
58	50	36	57
48	56	40	52

   Perform the analysis of variance and test the hypothesis of equal yield per acre for all varieties. Use $\alpha = 0.05$.

2. Derive the computational forms for SST, SSTR, and SSE given in section 10.1.1.

3. To show that SST = SSTR + SSE, we demonstrated in section 10.1.1, that

$$\sum_{i=1}^{k} (\overline{X}_i - \overline{X}) \sum_{j=1}^{n} (X_{ij} - \overline{X}_i) = 0$$

   Show that this is true for the data of exercise 1.

4. For exercise 1 write out the mathematical model and state any assumptions for each term of the model. Consider that the model is (i) fixed, and (ii) random.

5. For the data of exercise 1 calculate a 95% confidence interval for (i) μ_2, (ii) $\mu_1 - \mu_2$.

6. Suppose we wish to use orthogonal contrasts for the data of exercise 1 to compare $3\mu_1 - \mu_2 - \mu_3 - \mu_4$, $2\mu_2 - \mu_3 - \mu_4$, and $\mu_3 - \mu_4$. Find the appropriate sum of squares and mean squares. Test the appropriate hypotheses at $\alpha = 0.05$.

7. For the data of exercise 1 use Tukey's test with $\alpha = 0.05$ and Duncan's test with $\alpha_p = 0.05$ to separate the means.

8. Suppose that variety A has 5 observations. Two additional ones are 56 and 47. All other data remain the same. Use

the ideas of section 10.1.8 to test the hypothesis of equal yields with $\alpha = 0.05$. Then set a 95% confidence interval for μ_1 and for $\mu_1 - \mu_2$. Finally, use Duncan's test to separate the means.

9. Assuming a random effect model for the data of exercise 1, obtain unbiased estimates of σ^2 and σ_τ^2.

10. For the data of exercise 1 test that the variances are homogeneous at $\alpha = 0.05$, using (i) Hartley's test, and (ii) Bartlett's test. What are the chief disadvantages of the latter?

11.

Nonparametric Methods

Most of the parametric statistical techniques described in this book are optimal under certain standard assumptions. Some of these assumptions relate to the accuracy and method of obtaining the measurements while the others are concerned with the nature of data. For example, some important assumptions are concerned with the normality, independence, and homogeneity of variances among the observations. The scales of measurement and the method of obtaining the observations also affect the method of analysis. As described in chapter 1, there are four commonly used scales of measurements: nominal, ordinal, interval, and ratio. Interval and ratio scales are stronger and more useful than the nominal and ordinal scales because they have better communicative properties and the data obtained lend themselves easily to algebraic manipulations. Parametric methods assume that the measurements have at least the strength of the interval scale.

Nonparametric methods are concerned with the treatment of standard statistical problems when the assumptions of normality are replaced by general assumptions concerning the distribution function. Nonparametric methods are sometimes known as distribution-free methods. However, the observations certainly do have a distribution with parameters, although we are free from the restrictions and assumptions regarding the form of that distribution. Frequently the variables are just assumed to come from a continuous distribution.

Nonparametric methods offer certain advantages in that they require few assumptions, are simple to compute, especially when the samples are relatively small, and cope with observations that have less strength than the interval scale. In most cases they provide exact probabilities (except in the case of large samples, where excellent approximations are available) and the accuracy of the probability statement does not depend on the shape of the population distribution. Nonparametric tests are also available to analyze data that have only the strength of the ranking scale.

In spite of all the advantages of nonparametric tests, parametric tests are more powerful and accurate; Whenever all the assumptions of the parametric tests are met, they should be used in preference to nonparametric tests.

There are many nonparametric tests available, but we shall discuss only the following important and commonly used tests: Wilcoxon signed rank test, Wilcoxon rank sum test, Kruskal-Wallis H-test, Friedman χ_r^2-test, Spearman's rank correlation coefficient, McNemar chi-square test for related samples, and one-sample run test for randomness.

11.1 Wilcoxon Signed Rank Test

The Wilcoxon signed rank test is designed to investigate whether the two populations from which the samples are drawn have the same median. This test is a nonparametric counterpart to the parametric pairing design test in chapter 7 which assumes that the sampling distribution of differences of paired values is normal and that the measurements have at least the strength of the interval scale. The Wilcoxon signed rank test does not assume normality of population differences and can be used when the measurements have an underlying continuity and the strength of only the ranking scale.

Consider the data consisting of n pairs of observations (x_i,y_i), i = 1, 2, . . . , n. The observed differences $d_i = x_i - y_i$ are ranked in increasing order of magnitude and the ranks are given the signs of the corresponding differences. (Tied differences are given the average of the ranks that would have been given if no ties were present.) It is assumed that the differences $d_1, d_2, \ldots, d_n$ are mutually independent and come from continuous populations that are symmetric about 0.

Under the null hypothesis that the distribution of differences has a 0 median, we would expect the sum of the positive ranks to be approximately equal to the absolute value of the sum of the negative ranks. Let T be the smaller sum of the two absolute rank sums. We compare T with the critical rank sum corresponding values given in Appendix Table 15 for specified n (the number of pairs that gives a non-0 difference). For example, if N = 10 and we want a two-sided test at a 0.05 level of significance, we reject the null hypothesis if T is less than 8 or greater than 47. Note that a difference of 0 will occur if the paired measurements have the same value. Such paired measurements are dropped from the analysis. If the calculated T-value is less than or equal to the corresponding table T-value, the null hypothesis is rejected at the specified level of significance; otherwise it is accepted.

For a large sample size ($n > 25$), T is approximately normally distributed under the null hypothesis with

$$\text{Mean} = \mu_T = \frac{n(n+1)}{4},$$

and

$$\text{standard deviation} = \sigma_T = \sqrt{\frac{n(n+1)(2n+1)}{24}}$$

Therefore

$$z = \frac{T - \mu_T}{\sigma_T}$$

is approximately normally distributed with 0 mean and unit variance. Thus (using Appendix Table 3), the testing procedure is carried out as described in chapters 6 and 7.

Example 11.1 Ten calves were selected and given a special diet. Their weights, before they were subjected to the new diet, and after a lapse of 4 weeks, are recorded here.

Animal number	Weight (lb) Before	After
1	100	113
2	97	105
3	106	112
4	90	94
5	115	113
6	103	106
7	99	103
8	87	92
9	101	98
10	108	109

Test whether there has been any significant weight gain as a result of the special diet.

The differences and signed ranks are as follows:

d_i	8	3	6	4	−2	3	4	5	−3	1
Signed rank	10	4	9	6.5	−2	4	6.5	8	−4	1

$$\begin{aligned} T &= \text{total of those ranks with less frequent signs} \\ &= 6 \end{aligned}$$

From Appendix Table 15 the tabulated value of T for n = 10 at the 0.05 (two-sided) level of significance is 8. Since the calculated value of T(6) is less than the critical value, we conclude that the null hypothesis, that there is no effect of the special diet, is rejected at the 5% level of significance.

Although n is not large, we can try to apply the normal approximation. We have

$$\text{Mean} = \mu_T = \frac{n(n+1)}{4} = \frac{10 \times 11}{4} = 27.50$$

$$\text{SD} \quad = \sigma_T = \sqrt{\frac{n(n+1)(2n+1)}{24}}$$

$$= \sqrt{\frac{10 \times 11 \times 21}{24}} = 9.81$$

Thus,
$$z = \frac{T - \mu_T}{\sigma_T} = \frac{6 - 27.50}{9.81} = -2.19$$

Since the calculated $|z| = 2.19$ is greater than the 5% (two-sided) tabulated value of $Z = 1.96$, we reject the hypothesis that there is no effect of the special diet.

Note: We can also use the following formulas to test the significance of the calculated T value:

1. For a 5% level of significance calculate

$$T_{.05} = \frac{n^2 - 7n + 10}{5}$$

   If the calculated T is less than or equal to $T_{.05}$, the null hypothesis is rejected.

2. For a 1% level of significance calculate

$$T_{.01} = \frac{11n^2}{60} + 5 - 2n$$

   If calculated $T \leq T_{.01}$, reject the null hypothesis at 1% significance level.

Example 11.2 Two types of apple pie fillings, made by two different food processing companies (A and B), were used to make apple pies. Seven panelists were selected and given one sample from each type of apple filling. They scored the pies on a nine-point scale (1 = like extremely, 2 = like very much, 3 = like moderately, 4 = like slightly, 5 = neither like nor dislike, 6 = dislike slightly, 7 = dislike moderately, 8 = dislike very much, 9 = dislike extremely).

Test the following results to see if there is any difference between the two types of apple pie fillings in terms of preference:

Panelists		1	2	3	4	5	6	7
Type of apple pie filling	A	2	2	4	2	3	3	2
	B	3	5	3	9	1	7	6

The differences and signed ranks are as follows:

d_i	−1	−3	1	−7	2	−4	−5
Signed rank	−1.5	−4	1.5	−7	+3	−5	−6

$$T = \text{total of those ranks with less frequent signs}$$
$$= 4.5$$

The tabulated value of T for n = 7 at the 5% level of significance from Appendix Table 15 is $T_{.05} = 2$. Since the calculated T = 4.5 is greater than the table value of T = 2, we have no reason to reject the hypothesis that the two types of apple fillings are equally preferable.

11.2 Wilcoxon Rank Sum Test

The Wilcoxon signed rank test is applicable when the observations are paired. The Wilcoxon rank sum test, also known as Wilcoxon test for independent samples or Mann-Whitney U test, may be used to test whether two independent groups have been drawn from the same population. This powerful nonparametric test is an alternative and counterpart to the parametric t-test. The appropriate null hypothesis, in this case, is that the two independent samples come from identical populations. The alternative is that one population is shifted to the right (or left) or that the populations differ only in location.

We take independent samples of sizes n_1 and n_2 from the two populations and rank the combined sample of size $n = n_1 + n_2$, giving the smallest observation the rank of 1, and the largest the rank $n_1 + n_2$. (Ties are taken care of by giving them the average of the ranks that would have been assigned if no ties were present.) If the two populations are identical, we expect that the samples will intermingle in a regular manner. On the other hand, if there is any sizable difference between the location parameters, most of the lower ranks will be occupied by the observations from one sample while most of the higher ranks will be occupied by the observations from the other sample.

The test statistic U is the sum of the ranks for the sample of size n_1. This value is compared with the corresponding table value from Appendix Table 16 for the set (n_1,n_2) for a particular level of significance. If the calculated U value is less than the value obtained from Appendix Table 16, we reject the null hypothesis that the samples have come from the same population.

For large values of n_1 and n_2, U is approximately normally distributed with

$$\text{Mean} = \mu_U = \frac{n_1 n_2}{2}$$

$$\text{and standard deviation} = \sigma_U = \sqrt{\frac{n_1 n_2(n_1 + n_2 + 1)}{12}}$$

Thus,

$$z = \frac{U - \mu_U}{\sigma_U}$$

is approximately normal with mean 0 and unit variance.

Example 11.3 The following data give the results in percentage germination of seed samples by two different methods used by a laboratory in a germination trial. Is there a significant difference (5% level) between the two methods of germination?
Method A: 89, 85, 87, 90, 92, 88
Method B: 93, 94, 95, 89, 90, 86

For these data, the ranks are determined as follows:

Method	A	B	A	A	A	B	A	B	A	B	B	B
Percentage of germination	85	86	87	88	89	89	90	90	92	93	94	95
Rank	1	2	3	4	5.5	5.5	7.5	7.5	9	10	11	12

$$\begin{aligned} U &= \text{sum of ranks of sample size } n_1 \text{ (method A)} \\ &= 1 + 3 + 4 + 5.5 + 7.5 + 9 \\ &= 30 \end{aligned}$$

From Appendix Table 16, the critical value of U for $(n_1, n_2) =$ (6,6) is 26. Since the calculated value of U = 30 is greater than the table value of U = 26, we have no reason to reject the hypothesis of similarity of the results by two different methods.

If we apply the normal approximation, we get

$$\text{Mean} = \mu_U = \frac{n_1 n_2}{2} = \frac{6 \times 6}{2} = 18$$

$$\text{SD} = \sigma_U = \sqrt{\frac{n_1 n_2 (n_1 + n_2 + 1)}{12}}$$

$$= \sqrt{\frac{6 \times 6(6 + 6 + 1)}{12}}$$

$$= 10.54$$

Thus,

$$z = \frac{U - \mu_U}{\sigma_U} = \frac{30 - 18}{10.54} = 1.14$$

which is not significant at the 5% level of significance ($Z_{.05} = 1.96$). Therefore we arrive at the same conclusions as previously, that is, there is no difference between the two germination methods.

11.3 Kruskal-Wallis H-Test

When the measurements for the parametric group comparison test and completely randomized design are not met, the nonparametric counterpart Kruskal-Wallis H-test is used. The H-test does not assume that the populations are normally distributed and carries out the equivalent of one-way analysis of variance to test the hypothesis that the K samples have come from identical continuous populations.

Let n_i be the size of the *i*th sample and $N = n_1 + n_2 + \ldots + n_K$ be the total number of independent observations in the K samples. For the H-test each of the N observations is replaced by a rank, that is, all of the scores from all of the K samples combined are ranked in a single series. After this has been done the sum of the ranks in each sample is found. We denote by R_i the sum of the ranks in the *i*th sample. Then the Kruskal-Wallis H-test is defined as

$$H = \frac{12}{N(N + 1)} \sum_{i=1}^{K} \frac{R_i^2}{n_i} - 3(N + 1)$$

If the null hypothesis of K samples being from the same or identical populations is true, H is approximately distributed as chi-square with K − 1 degrees of freedom. Thus, the calculated value of H is compared with the chi-square value for K − 1 degrees of freedom at a specified level of significance (using Appendix Table 5). If the calculated value of H is greater than the corresponding table chi-square value, we reject the null hypothesis and conclude that the K samples have not come from the same population.

11.3.1 Correction for Ties

If two or more measurements occur with the same value, a correction for ties can be made. The correction factor (CF) is given by

$$CF = 1 - \frac{\Sigma M}{N^3 - N}$$

where $M = m^3 - m$, m being the number of tied observations in a tied group of scores.

The corrected H is given by

$$\text{Corrected H} = \frac{H}{CF}$$

The correction factor is almost negligible when the number of ties is small. The corrected H-value is always larger than the H-value. Therefore, if we are able to reject the null hypothesis without correcting the H, we will certainly be rejecting with corrected H.

Example 11.4 The following data give liver weight (gm) of female rats who have been subjected to hormone treatment. The categories A, B, and C represent the liver weight before treatment, after 24 hours of treatment, and after 96 hours of treatment, respectively. Using Kruskal-Wallis H-test, determine whether there is any significant gain in liver weight as a result of hormonal treatment.

Liver Weight in Female Rats

A = Before treatment	B = After 24 hours of treatments	C = After 96 hours of treatments
7.865	9.687	11.342
8.684	10.023	11.542
8.213	9.781	12.013
7.561	10.562	11.543
8.010	11.010	10.876
8.532	9.854	
7.048		
8.102		
6.850		
8.324		

Here,

$$K = \text{number of samples} = 3$$

$$N = \sum_{i=1}^{K} n_i = n_1 + n_2 + n_3$$

$$= 10 + 6 + 5 = 21$$

All the N observations are ranked in a single series and the sums of the ranks within sample (R_i) are calculated (see table 11.1).

Kruskal-Wallis H-test may be calculated as follows:

$$H = \frac{12}{N(N + 1)}\left[\frac{R_1^2}{n_1} + \frac{R_2^2}{n_2} + \frac{R_3^2}{n_3}\right] - 3(N + 1)$$

$$= \frac{12}{21 \times 22}\left[\frac{(55)^2}{10} + \frac{(82)^2}{6} + \frac{(94)^2}{5}\right] - 3 \times 22$$

$$= 16.854$$

Table 11.1 *Ranked observations with sums from example 11.4.*

A	Rank	B	Rank	C	Rank
7.865	4	9.687	11	11.342	18
8.684	10	10.023	14	11.542	19
8.213	7	9.781	12	12.013	21
7.561	3	10.562	15	11.543	20
8.010	5	11.010	17	10.876	16
8.532	9	9.854	13		
7.048	2				
8.102	6				
6.850	1				
8.324	8				
R_i	55		82		94

The chi-square value from Appendix Table 5 for $K - 1 = 3 - 1 = 2$ degrees of freedom at the 5% level of significance is $\chi^2_{0.95} = 5.991$. Since the calculated H-value of 16.854 is greater than $\chi^2_{0.95} = 5.991$, we reject the null hypothesis of no treatment effect and conclude that there has been a significant increase in liver weight due to hormone treatment.

11.4 Friedman's χ^2_r-Test

Friedman's χ^2_r-test is a nonparametric counterpart of randomized block design. It makes no assumption about the normality of the data and can cope with measurements that have only the strength of the ordinal scale. The null hypothesis tested is that measurements in each block (row) are from the same population. The data are set in a block design with K treatments each having n blocks. The measurements are ranked in each block over the treatments. When this has been done for each block, the ranks are summed in each column or treatment. The Friedman's χ^2_r-test statistic is given by

$$\chi^2_r = \frac{12}{nK(K + 1)} \sum_{i=1}^{K} R_i^2 - 3n(K + 1)$$

where

R_i = sum of the ranks in the *i*th column

K = number of treatments

n = number of blocks (rows)

The upper significance level values of χ_r^2 are given in Appendix Tables 17.1 and 17.2 for moderate values of K and n. For n and K outside this range, χ_r^2 has an approximate chi-square distribution with K − 1 degrees of freedom. Therefore, the calculated value of χ_r^2 is compared with the tabulated χ^2-values (Appendix Table 5) for K − 1 degrees of freedom for a given level of significance. If the calculated χ_r^2-value is greater than the table χ^2-value, we reject the hypothesis of population equality.

When K = 2 and n is relatively small, Wilcoxon signed rank test is preferred over Friedman's χ_r^2 test.

Example 11.5 Five different labeling machines were tested in a food processing plant. The data record the number of defectively labeled cans in a 5-day production run.

		Machine				
		A	B	C	D	E
	1	5	9	3	6	7
	2	6	7	2	4	9
Day	3	7	11	1	5	8
	4	4	8	2	3	10
	5	3	6	1	4	7

We shall use Friedman's χ_r^2 test with $\alpha = 0.05$ to see if there is any difference between machines. The calculations for ranking the observations are given in table 11.2.

Thus,

$$\chi_r^2 = \frac{12}{nK(K+1)} \sum_{i=1}^{K} R_i^2 - 3n(K+1)$$

$$= \frac{12}{5 \times 5 \times 6} [(13)^2 + (22)^2 + (5)^2 + (12)^2 + (23)^2] - 3 \times 5 \times 6$$

$$= 18.12$$

Table 11.2 *Ranking calculations for example 11.5.*

		Machine									
		A	Rank A	B	Rank B	C	Rank C	D	Rank D	E	Rank E
	1	5	2	9	5	3	1	6	3	7	4
	2	6	3	7	4	2	1	4	2	9	5
Day	3	7	3	11	5	1	1	5	2	8	4
	4	4	3	8	4	2	1	3	2	10	5
	5	3	2	6	4	1	1	4	3	7	5
		$R_1 = 13$		$R_2 = 22$		$R_3 = 5$		$R_4 = 12$		$R_5 = 23$	

The value of $\chi^2_{.05}$ for K − 1 degrees of freedom = 5 − 1 = 4 from Appendix Table 5 is 9.49. Since the calculated value of χ^2_r = 18.12 is less than the table value of $\chi^2_{.05}$ = 9.49, the result is significant and we reject the null hypothesis of no difference between the machines.

11.5 Spearman's Rank Correlation Coefficient

The coefficient of correlation discussed in chapter 9 assumes that the two characters are definitely measurable, that is, the measurements have at least the strength of an interval scale. It also assumes that the measurements are drawn from a bivariate normal population. When such parametric assumptions cannot be satisfied, we can use the nonparametric Spearman's rank correlation coefficient (r_s) to measure the relationship between two variables X and Y.

Suppose we have n paired observations for the variables X and Y. The observations in each variable are ranked separately, i.e., columnwise, and the differences between the ranks appearing in each row are obtained. When there are no ties, r_s is calculated as follows:

$$r_s = \frac{6 \Sigma d^2}{n^3 - n}$$

11.5.1 Correction for Ties

If two or more measurements with the same value occur in each of the two variables X_i and Y_i, a correction for ties can be made. We calculate

$L = \Sigma (\ell^3 - \ell)$, ℓ being the number of measurements with the same value within the variable X

$M = \Sigma (m^3 - m)$, m being the number of measurements with the same value within the variable Y

Thus, the r_s value corrected for ties is given by

$$r_s = \frac{n^3 - n - 6\,\Sigma\, d^2 - \frac{1}{2}(L + M)}{\sqrt{(n^3 - n - L)(n^3 - n - M)}}$$

If there are no tied ranks, i.e., L = M = 0, this formula for r_s reduces to the one given for the case of no tied ranks. Also, when the proportion of ties is small, the correction need not be applied.

Like the interclass correlation coefficient r described in chapter 9, the Spearman's coefficient r_s lies between −1 and +1. Thus, $r_s = +1$ for the case of perfect agreement and $r_s = -1$ for the case of perfect disagreement between the two series of ranks. The test of the null hypothesis that the two variables under study are not associated in the population can be carried out using the t-distribution. To do this, calculate

$$t = \frac{r_s}{\sqrt{\frac{1 - r_s^2}{n - 2}}}$$

and compare this calculated value of t with table value of t (Appendix Table 4) for n − 2 degrees of freedom for a specified level of significance. If the calculated value of t is greater than the table t-value, we reject the hypothesis of no association between the two groups.

Example 11.6 In an apple packing plant two inspectors ranked 10 workers according to their efficiency as follows:

Worker	Inspector A	Inspector B
1	4	3
2	8	9
3	6	6
4	7	5
5	1	1
6	3	2
7	2	4
8	5	7
9	10	8
10	9	10

Calculate the Spearman's rank correlation coefficient r_s, and test the hypothesis of no similarity between the opinion of two inspectors.

Since the data are already in ranks, we calculate the difference d between the ranks of two inspectors for each worker as follows:

$$\begin{aligned} d: &\ 4 - 3, 8 - 9, 6 - 6, 7 - 5, 1 - 1, 3 - 2, \\ &\ 2 - 4, 5 - 7, 10 - 8, 9 - 10 \\ &= 1, -1, 0, 2, 0, 1, -2, -2, 2, -1 \end{aligned}$$

From this we get

$$\Sigma d^2 = 20$$

Thus,

$$r_s = \frac{6 \Sigma d^2}{n^3 - n}$$

$$= 1 - \frac{6 \times 20}{(10)^3 - 10} = 0.88$$

To test the hypothesis of no association between the scores of the two inspectors, we calculate

$$t = \frac{r_s}{\sqrt{\dfrac{1 - r_s^2}{n - 2}}}$$

$$= \frac{0.88}{\sqrt{\frac{1 - (0.88)^2}{10 - 2}}} = 5.08$$

For a two-sided test the table value of t for 10 − 2 = 8 degrees of freedom from Appendix Table 4 is 2.306. Since the calculated value of t = 5.08 is greater than the table value of t = 2.306, we reject the null hypothesis of no association and conclude that the opinion of the two inspectors with regard to the efficiency of the workers shows great similarity.

Example 11.7 Consider the following 12 paired measurements giving the time (in minutes) taken by two workers to finish a job. The data and the calculations for the rank differences are given in table 11.3.

Table 11.3 *Rank differences for example 11.7.*

Pair	Worker A	Rank	Worker B	Rank	d	d^2
1	32	9	29	2.5	+6.5	42.25
2	36	12	34	10	+2.0	4.00
3	31	7	30	4.5	+2.5	6.25
4	30	4.5	32	8	−3.5	12.25
5	29	2	36	12	−10.0	100.00
6	30	4.5	31	6.5	−2.0	4.00
7	32	9	29	2.5	+6.5	42.25
8	35	11	30	4.5	+6.5	42.25
9	30	4.5	34	10	−5.5	30.25
10	28	1	27	1	0	0.00
11	30	4.5	31	6.5	−2.5	6.25
12	32	9	34	10	−1.0	1.00
						$\Sigma d^2 = 290.75$

Thus,

$$r_s = \frac{6 \Sigma d^2}{n^3 - n}$$

$$= 1 - \frac{6 \times 290.75}{(12)^3 - 12}$$

$$= -0.02$$

Since this value of $r_s = -0.02$ is very close to 0, we accept the hypothesis of no correlation between the two sets of measurements. As there are several tied observations, we calculate the r_s value using the formula for tied observation. We calculate the quantities L and M as follows:

For X_i (worker A) values, there are
4 measurements of 30
3 measurements of 32

Therefore,

$$\begin{aligned} L &= \Sigma (\ell^3 - \ell) \\ &= (4^3 - 4) + (3^3 + 3) \\ &= 84 \end{aligned}$$

Similarly, we can calculate M for the Y_i (worker B) values as

$$\begin{aligned} M &= \Sigma (m^3 - m) \\ &= (2^3 - 2) + (2^3 - 2) + (2^3 - 2) + (3^3 - 3) \\ &= 42 \end{aligned}$$

Therefore, the corrected r_s is given as

$$r_s = \frac{n^3 - n - 6 \Sigma d^2 - \frac{1}{2}(L + M)}{\sqrt{(n^3 - n - L)(n^3 - n - M)}}$$

$$= \frac{12^3 - 12 - 6 \times 290.75 - \frac{1}{2}(84 + 42)}{\sqrt{(12^3 - 12 - 84)(12^3 - 12 - 42)}}$$

$$= -0.07$$

For the test of null hypothesis, we calculate

$$t = \frac{r_s}{\sqrt{\frac{1 - r_s^2}{n - 2}}}$$

$$= \frac{-0.07}{\sqrt{\frac{1 - (-0.07)^2}{12 - 2}}} = -0.222$$

The table $t_{.05}$ value for $12 - 2 = 10$ degrees of freedom from Appendix Table 4 is 2.228. Since the calculated value of $t = 0.07$ is less than the table value, we accept the hypothesis of no correlation between two sets of measurements.

11.6 McNemar Chi-Square Test for Related Samples

In chapter 8 we discussed the method of analyzing independent samples in a contingency table. Here we shall consider the McNemar test for analyzing 2×2 contingency tables when the samples are related. An example is when two vaccines are compared for their effectiveness by administering one vaccine "before" on several animals, and the other "after" on the same animals and recording their response. The general features of such a 2×2 contingency table are shown as follows:

			Vaccine I	
			Reaction	
			−	+
Vaccine II	Reaction	−	a	b
		+	c	d

Notice that cases that show changes between the first and second response appear in cells b and c. Only cells b and c contribute information leading to the acceptance or rejection of the null hypothesis. The McNemar test for testing the significance of changes is given by

$$\chi^2 = \frac{(b - c)^2}{b + c} \text{ with 1 degree of freedom}$$

Example 11.8 Two experimental vaccines were tested on 60 animals. Each animal was given vaccine I and the response noted. The response was recorded as "+" if the symptoms were relieved and "−" otherwise. After a period of several months when it was felt that there is no carry-over effect from vaccine I, vaccine II was administered in the same manner on the same 60 animals. The data are presented in the following 2 × 2 contingency table.

		Vaccine I −	Vaccine I +
Vaccine II	−	8	20
	+	18	14

Using the McNemar chi-square test, we have

$$\chi^2 = \frac{(20 - 18)^2}{20 + 18} = 0.11 \text{ with 1 degree of freedom}$$

The $\chi^2_{.05}$ value for 1 degree of freedom from Appendix Table 5 is 3.841. Since the calculated value of $\chi^2 = 0.11$ is less than the table value of $\chi^2 = 3.841$, we accept the hypothesis of no difference in effectiveness between the two vaccines.

11.7 One-Sample Run Test for Randomness

Estimation of population parameters is generally done from sample results. It is, therefore, desirable to select a random representative sample. To test the hypothesis that a sample is random, we use the order in which the observations were originally obtained. The technique used here is based on the theory of runs. A *run* is defined as a succession of identical symbols followed and preceded by different symbols or by no symbols at all. For example, suppose a series of plus or minus scores occurred in this order:

$$- + + - - - + - - + + + - - - - - + + - +$$

We have a run of 1 minus, followed by a run of 2 pluses, then a run of 3 minuses, and so on. The total number of runs r is equal to 10. Note the total sample size N = 20, the number of pluses n_1 = 9, and the number of minuses n_2 = 11. In sampling inspection we have runs of defective and nondefective items; in tossing a coin we have a run of heads and tails. The total number of runs in a sample of a given size indicates whether a sample is random or not. Too few or too many runs may indicate a time trend or systematic short-term cyclical fluctuations in the observations.

Let N be the total number of items in the sample of which n_1 are of one kind and n_2 are of the other kind. n_1 and n_2 can be good or bad items in a sampling inspection of items from a lot or they may be the measurements below the sample median (−) and the measurements above the sample median (+).

For the one-sample run test we have to find the number of runs r in the whole group of N observations. For small sample size the test procedure is carried out by using the critical values of the distribution of r given in Appendix Table 18. The table gives the 5% two-sided values with the smaller integer being the left-tail critical value and the larger integer being the right-tail critical value. To test the null hypothesis of randomness at $\alpha = 0.05$, we count the number of runs in the sample and compare it with table values (Appendix Table 18). If the number of runs r in a given sample falls between the two values given in the table, we accept the null hypothesis of randomness.

For large sample sizes n_1 or $n_2 > 20$, r is normally distributed with

$$\text{Mean} = \mu_r = \frac{2n_1 n_2}{n_1 + n_2} + 1$$

$$\text{and standard deviation} = \sigma_r = \sqrt{\frac{2n_1 n_2(2n_1 n_2 - n_1 - n_2)}{(n_1 + n_2)^2 (n_1 + n_2 - 1)}}$$

Thus,

$$z = \frac{r - \mu_r}{\sigma_r}$$

is approximately normally distributed with 0 mean and unit variance.

Example 11.9 A labeling machine in a food processing plant produces a series of defective (−) and nondefective (+) labels. One such series of events is recorded here:

+ + + + − + + + + + − − + + + − + + + + + + + − + +

N = 25, n_1 = 20, n_2 = 9, and r = 9.

The 5% values for (n_1,n_2) = (20,5) (from Appendix Table 18) are (5,12). Since our calculated value of r = 9 falls between the two table values, we accept the hypothesis of randomness of events.

Exercises

1. In a seed testing laboratory, the experimenter carried out a seed certification trial with a particular variety of grass seed produced by two different manufacturing companies, A and B. One seed was taken from each of the samples provided by company A and B and grown in the same population. The data record the number of hours taken by the seed to emerge after planting.

	Hours taken to emerge	
Seed number	Company A	Company B
1	47	42
2	49	48
3	46	49
4	45	47
5	43	45
6	48	42
7	47	45
8	51	49
9	50	48
10	49	46

   Using Wilcoxon signed rank test, investigate the difference between the emergence time for the seeds produced by the two companies.

2. The following data give scores on a five-point scale by 5 panelists for canned beans produced by two processing plants, A and B. The five-point scale is given by: 1 = like very much, 2 = like, 3 = neither like nor dislike, 4 = dislike, 5 = dislike very much.

Panelist number	Scores for beans produced by Plant A	Plant B
1	3	1
2	4	1
3	4	2
4	5	2
5	3	2

Use Wilcoxon signed rank test to test the hypothesis of no difference between the scores for the two types of beans.

3. Ten samples of skim milk powder were taken from a lot. Each sample was divided into two portions and analyzed by two methods, A and B, for moisture content. The following data record the percentage of moisture for the 10 samples.

Sample number		1	2	3	4	5	6	7	8	9	10
Percent moisture	Method A	3.5	4.2	3.8	3.0	3.5	2.9	3.1	3.5	3.6	3.0
	Method B	3.2	3.7	3.9	3.2	3.0	2.8	3.1	2.9	3.4	3.2

Test the difference between the two methods using Wilcoxon signed rank test.

4. The following data consist of scores given by 10 judges for juiciness of hamburgers with three additives. The three additives were (a) salt (b) salt, egg, (c) salt, egg, bread. The scores were given on the following scale: 1 = very juicy, 2 = juicy, 3 = moderately juicy, 4 = not juicy

Judge number	Hamburgers with Salt	Salt, Egg	Salt, Egg, Bread
1	1	2	1
2	1	2	2
3	3	1	2
4	2	3	3
5	1	1	3
6	3	1	2
7	1	1	2
8	1	3	3
9	2	2	1
10	1	1	2

Using Kruskal-Wallis H test, investigate the difference in preference between three types of hamburger additives.

5. Consider exercise 1, chapter 3. Investigate the difference between laboratories by analyzing the data using Kruskal-Wallis H-test.

6. The following are the number of mutilated cans produced by two machines in a cannery.

   Machine I: 21, 24, 27, 31, 26, 19, 23, 30
   Machine II: 23, 29, 25, 17, 26

   Using the Wilcoxon Rank sum test for $\alpha = 0.05$, test the null hypothesis that the samples are drawn from identical populations.

7. Consider the data on yield of tomato plants given two fertilizers A and B from exercise 5, chapter 3. Analyze the data using Wilcoxon rank sum test to investigate the difference between the two fertilizers.

8. The sampling inspection of potatoes from a lot by two inspectors produced the following results in terms of percent defective for the characteristic of hollow heart.

   Inspector A: 4.2 5.3 4.9 6.8
   Inspector B: 5.4 9.2 6.1 5.1 4.8 5.5 6.0

   Apply Wilcoxon rank sum test to determine the difference between the results of the two inspectors.

9. Analyze the sensory evaluation data of exercise 4 using the Friedman's χ_r^2-test.

10. The following data give the yield in pounds of potatoes in a fertilizer experiment:

	Treatment			
Block	A	B	C	D
1	187	291	295	305
2	189	283	301	310
3	194	277	273	300
4	178	269	289	311

Test the difference between treatments using Friedman's χ_r^2-test.

11. Using Friedman's χ_r^2-test, analyze the following data giving the yield of grain in grams per 15-foot row (each row 8 inches apart) in an experiment with 5 varieties of barley.

	Blocks			
Variety	1	2	3	4
1	300	420	341	349
2	325	402	362	363
3	401	415	382	373
4	375	395	315	400
5	390	389	371	360

12. In a feeding trial experiment with pigs, the gain in weight in pounds was recorded as follows:

Batch			
1	2	3	4
200	152	121	182
197	141	134	191
189	139	129	199
175	160	130	179
201	159	135	186
167	132	140	198

Using Kruskal-Wallis H-test, determine whether there is any difference between the batches.

13. The following data give the yield of 3 strains of wheat planted in 5 randomized blocks.

	Blocks				
Strains	1	2	3	4	5
A	27	24	26	25	28
B	21	20	22	19	23
C	23	24	22	20	25

Using Friedman's χ_r^2-test, study the difference between the 3 strains.

14. In a dairy cattle show 2 judges gave the following scores out of a total of 100 points to 7 animals:

Animals	1	2	3	4	5	6	7
Judge A	89	83	92	75	84	80	87
Judge B	88	84	91	76	83	80	89

Using Spearman's rank correlation coefficient r_s, test the hypothesis of no correlation between the scores of the 2 judges.

15. Analyze the data of exercise 2, using Spearman's rank correlation coefficient.

16. The mean soil temperature and germination interval for winter wheat in 10 places are:

Mean soil temperature	58	41	45	39	43	44	53	56	44	49
Number of days	16	21	25	37	17	19	23	32	22	26

Investigate the degree of correlation between the two sets of observations using Spearman's rank correlation coefficient.

17. The following 2 × 2 contingency table gives data obtained during a clinical trial where two drugs, A and B, were administered and the response noted in terms of whether the drugs produced allergic reaction:

		Drug A	
		No Reaction	Reaction
Drug B	No reaction	6	14
	Reaction	32	11

Investigate the difference between the two drugs using McNemar chi-square test for related samples.

18. Test the randomness of the twenty observations given here by considering the sequence of positive and negative signs.
26, 23, 30, 8, 20, 62, 45, 6, 2, 9,
35, 11, 49, 2, 51, 33, 25, 41, 1, 60

19. In a sample size of 45 there are 19 runs above and below the median. Is the sample random?

20. In a pie-testing competition, two judges provided the following ranks for 6 pies:

	Competitor					
	1	2	3	4	5	6
Judge A	4	5	2	1	7	3
Judge B	5	7	3	6	4	2

Measure the association between the judgments of the two judges by using Spearman's rank correlation coefficient.

12.

Survey Sampling

In any investigation, whether a laboratory experiment, a sampling inspection of food products, an opinion poll, or any survey sampling situation, the primary question a researcher asks is, How large a sample should be selected and what sampling method should be used for selecting the sample?" This chapter and the next are devoted to some important sampling methods and sampling plans and the related estimation procedures.

For any statistical investigation, complete enumeration of the population is impractical, or impossible if the population is infinite. Also, if the units are destroyed in the course of inspection, 100% inspection, though possible, is not at all desirable. But even if the population is finite or the inspection is not destructive, 100% inspection is not used because it is time consuming, uneconomical, and creates fatigue and boredom for the inspection personnel. The results of a sample survey are generally more reliable than those obtained from a complete census because it is possible to determine the extent of sampling errors from a sample survey. However, there are certain limitations in the sampling process. For instance, the sampling units have to be drawn by using appropriate sampling techniques, the sample size has to be adequate, and trained and qualified personnel have to be employed for proper planning, execution, and analysis.

Sampling aims to estimate such population parameters as the mean or total from information contained in a sample. Sample characteristics are used to determine or estimate the population. The quantity and quality of information available from the sample depends heavily on the sample size and the method used to select the sample data. The amount of sampling to be done depends on the degree of accuracy desired by the experimenters and the availability of money to obtain that accuracy. If θ is the parameter of interest and $\hat{\theta}$ is an estimator of θ, we define the error of estimation as follows:

$$\text{Error of estimation} = |\theta - \hat{\theta}|$$

It is up to the experimenters to specify the boundaries and the probability of the error of estimation; they should specify that in repeated sampling, the error of estimation is required to be less than some value e with a probability of $1 - \alpha$. In the case of normally distributed data, we have the error of estimation less than $1.96 \times \text{SE of } \hat{\theta}$. For most practical purposes we shall take $e = 2\sigma_{\hat{\theta}}$, and hence $1 - \alpha$ will be approximately 0.95 for most bell-shaped distributions. This can be stated as

$$P[|0 - \hat{\theta}| < 2\sigma_{\hat{\theta}}] = 0.95$$

12.1 Principles of Sample Surveys

Most sampling inquiries can be broadly classified into two groups: (i) inquiries that can be answered by conducting a carefully designed and controlled sampling experiment, and (ii) inquiries that can be answered by conducting a sample survey where the individuals cannot be subject to experimental control.

Three basic principles should be considered when designing a sample survey. They are regularity, validity, and optimization.

The principle of statistical regularity recommends the selection of a moderately large sample size, chosen at random from a lot or population. A larger sample will, on the average, provide better estimates of the population parameters than a small one.

The validity principle stresses the need for selecting representative samples that will provide reliable and valid estimates of the true population parameters, thereby providing objective interpretations of the results.

The principle of optimization takes into account two factors: efficiency and cost. Efficiency is measured by the inverse of the sampling variance of the estimator, while the cost is measured by the expenditure incurred. The principle of optimization strives to achieve maximum possible efficiency with minimum cost or with a given level of cost. The emphasis is on choosing an optimum design to cut losses due to errors of estimation and cost of sampling.

12.2 Steps of a Sample Survey

There are several steps involved in conducting a sample survey, though each particular situation dictates its own important requisites. The four main stages in conducting a survey are planning, execution, analysis, and reporting. Some of the elements and functions embodied within these four stages are described here:

1. Objective setting: The primary task is to define the objectives of the survey in the most clear, concise, and concrete terms. The objectives should consider short-term as well as long-term implications. They should be commensurate with available resources and target dates.

2. Defining the population: Proper care should be taken to define, in clear and unambiguous terms, the population or the aggregate of individuals or things from which the sample is to be selected. The information obtained from the sample will be valid and more widely applicable if it is taken from a properly defined population.

3. Defining the sampling frame and unit: A *sampling unit* is an unambiguous, nonoverlapping collection of elements from the population and a *frame* is a list of sampling units. Since a frame serves as a guide to the population to be covered and also determines the structure of the sample survey, it should be carefully and precisely defined. A frame can be a list, map, or other acceptable material. Sampling units should cover the entire population, and every element of the population should belong to one and only one sampling unit.

4. Data collection: Data should be collected keeping in view the objective of the survey. Efforts should be directed to collecting essential, precise, and relevant information. This can be accomplished by deciding in advance on the mode of presentation (e.g., tables, charts) of data that the survey should produce and then gathering the requisite information accordingly. After a decision has been made on the nature of the data to be collected, the most important step is to construct a questionnaire or schedule of inquiry. This requires careful attention, special skill, and familiarity with the subject matter. The inquiry should be clear, unambiguous, and to the point. The respondent should be given suitable and detailed instructions for the completion of the inquiry and very little room should be left for guessing.

5. Method of data collection: The most commonly used methods of data collection are personal interviews, telephone interviews, mailed questionnaires, and direct observation. In the interview method, whether personal or by telephone, the investigator asks the questions one by one and fills up the schedule on the basis of the information supplied by the individuals. In the third method the questionnaire is mailed to the individuals who fill it out on their own and return the completed information by mail to the investigating agency. Quite often the data cannot be collected due to nonresponse for many reasons such as an unavailable sampling unit or individual, or an individual who refuses to furnish the information. Properly trained personnel should be employed for interviews and follow-up action should be employed for the mailed questionnaire.

6. Sampling scheme: The next step in a sample survey is to specify the sampling design, i.e., the size of the sample, the method of selecting the sample, and the procedures for estimating population parameters. The subsequent sections of this chapter describe the various important procedures for designing the sample survey.

   Once the planning stage has been completed, the execution stage involves identification of the sampled individuals in the field and the filling in of the questionnaire. The analysis and reporting stages consist of the following steps: scrutinizing the data, tabulating the data, statistical analysis, reporting of results, and sorting out information for future surveys.

12.3 Sampling and Nonsampling Errors

In conducting a sample survey, two types of errors generally occur during the collection, processing, and analysis of data. They are sampling errors and nonsampling errors.

1. Sampling errors: Sampling errors occur as a natural consequence of the fact that only a portion of the population is used to estimate population parameters. Such errors are due primarily to the following reasons:
   (a) Bias due to faulty sampling method: Some errors can be introduced by the use of nonrandom sampling methods such as purposive sampling or grab sampling. The final estimates can be biased and misleading. This can be avoided by using probability sampling methods.
   (b) Bias due to faulty demarcation of sampling units: Sometimes defective designations or demarcation of sampling units can introduce errors. This is particularly significant in agricultural experiments in the field or in crop-cutting surveys. Such problems should be carefully studied during the planning stage and the procedure properly established and spelled out for the investigator.
   (c) Bias due to substitution: Such biases commonly occur when practical difficulties arise in enumerating a particular sampling unit or individual included in the random sample, and the investigator substitutes a conveniently available member of the population. Since the substituted individual may or may not possess the same characteristics as the unit originally intended for inclusion in the sample, serious biases can occur.
   (d) Errors may also occur due to improper choice of statistics used for estimating the population parameters. Inferences about the population drawn from such faulty statistics can be misleading and erroneous. In such

cases, selecting larger samples would result in a decrease of sampling error. The sampling error, in some cases, is inversely proportional to the square root of the sample size.

2. Nonsampling errors: Whereas sampling errors occur as a byproduct of the inductive inference made about the population on the basis of a sample, nonsampling errors can occur at any stage of planning or execution in both sample survey and complete enumeration. Although 100% inspection may be free from sampling errors, it is still subject to nonsampling errors. Nonsampling errors are likely to be more serious in a complete census than in a sample survey.

   Some commonly occurring nonsampling errors during the stages of observation, ascertainment, and processing of data include:

   (a) Errors due to poor planning: A sample survey should include a rigorous exercise of proper planning and organization and an explicit statement of objectives, specification of procedures, and definition of specifications. In the absence of proper planning, many nonsampling errors can occur due to lack of trained personnel, misinterpretation of the specifications, and faulty choice or selection of sampling units or sampling frame among other things.

   (b) Response errors: These errors are introduced as a result of the responses furnished by the respondent. They may be accidental or unintentional as when the respondent misunderstands a particular inquiry and furnishes improper information. On the other hand, the respondent may be intentionally providing false and misleading information.

   (c) Errors due to nonresponse: These errors may occur because of nonavailability of the sampling unit or because an investigator does not make a concerted effort to seek all sampling units.

   (d) Errors in coverage: If the directions for data collection are not clearly understood, the investigator may wrongly include certain units that are not to be included, or exclude certain units that are intended for inclusion in the sample.

(e) Last, many nonsampling errors can occur during compilation, tabulation, analysis, reporting, and publication of the data.

12.4 Types of Sampling

The method of sample selection is of primary importance to researchers because they have to find a method that will provide a highly reliable sample as well as conveniently fit the type or location of the population or lot. Sampling methods can be broadly classified as nonprobability sampling and probability sampling.

12.4.1 Nonprobability Sampling

Nonprobability sampling is sometimes referred to as subjective or purposive sampling. In this type of sampling, the investigator exercises discretion or judgment in deciding which items are to be included in the sample. Since no probabilities are assigned for the selection of sampling units, some items from the population have a better chance of inclusion than others. Thus, the sampling biases cannot be calculated and the accuracy of the estimates cannot be determined. Some of the most commonly used nonrandom sampling methods are described here:

1. Judgment sampling: In judgment sampling the investigators use their own judgment and experience regarding the population, lot, or sampling frame to decide which sampling units to select. Since sample selection is left exclusively to the discretion of the sampler and each sampler may have different criteria of judgment, an element of nonuniformity can be introduced into the system. Also, since the sampling is done without any probabilistic considerations, measurement of errors is difficult and proper statistical techniques cannot be applied for studying the precision of the estimates.

Although judgment sampling is nonrandom, the practical difficulties of a particular situation, best comprehended by the sampler, may suggest this method as the only practical and feasible method of selecting the sample. The employment of sound judgment is indispensable to any sampling procedure, be it random or nonrandom. If an investigator is highly skilled and experienced and knows the product or situation very well, judgment sampling provides more valuable information about the global objectives of the study than strictly structured and compartmentalized probabilistic methods. Consider, for example, the case of food inspectors carrying out sampling inspection of food or agricultural products for certification or compliance to standards and specifications. They have worked with the product for many years and have developed a fairly good idea about the quality of that product. The inspectors are familiar with the physical placements of the lots from which the samples are to be drawn, and also know the inherent cyclical trends or patterns of quality built into the production process. In such a case it is perhaps more profitable to carry out inspection through judgment sampling or obtain information through subjective monitoring than through random sampling. A random sample, or a sample over which the inspector has no control, may provide misleading information about the true state of quality control or results that are unrepresentative of the lot or population.

In summary, we should attempt to use probability sampling methods as much as possible and supplement the method with sound judgment and the benefit of our experience.

2. Quota sampling: This method consists in specifying in advance quotas of the samples to be selected from different sampling frames or subgroups of lots or populations. Sample selection is done by either judgment or random sampling. If the quotas are decided scientifically, and the sampling is done through probability sampling methods, the results can be reliable; otherwise this method can provide unreliable and inefficient results.

3. Restricted sampling: When a sample is taken from only a portion of the accessible lot or population, it is known as a *restricted sample*. The restriction can be natural, artificial, or purposely created for convenience. A restricted sample may not represent the population fully, and therefore may not provide meaningful generalizations for the entire population. In many situations concerning sampling inspection for quality control of food products, the physical placement of lots creates accessibility problems and inspections can only be carried out by taking convenient samples such as from a heavily loaded box car or heavily packed warehouse, or bulk sampling of bins of grapes.

4. Convenience sampling: When an investigator selects a sample in any convenient way, the process is termed *convenience sampling.* Other terms such as *chunk sampling, grab sampling,* and *haphazard sampling* can also be used. In chunk or grab sampling the investigator grabs a portion or chunk that is convenient; for example, the top layer in a box, units near the entrance, or first pallet. A haphazard sample is drawn in any way the investigator likes. It is difficult to apply statistical techniques to such types of sampling methods.

12.4.2 Probability Sampling

Probability sampling eliminates the discretion of the investigator with regard to which sampling units are to be included in the sample. The sample is selected scientifically in accordance with the principle of statistical sampling and probability. The sample is drawn by a random selection method and every individual in the population has a known probability of being included in the sample. The procedures for forming the estimates are known beforehand and the degree of precision of the estimates can be stipulated. Here we consider the three most important probability sampling methods: simple random sampling, systematic sampling, and stratified random sampling.

Simple Random Sampling

A simple random sample is selected from a lot or population using a random process and is the one method in which all elements in the lot have an equal and independent chance of being included in the sample.

An equal probability of selection is assigned to each unit of the population at the first draw. This implies an equal probability of selecting any unit from the available units at subsequent draws. Thus, if a sample of size n is to be drawn from a lot of size N, the probability of drawing any unit at the first draw is $\frac{1}{N}$. The probability of drawing any unit in the second draw (when the sampling is without replacement) from among the available (N − 1) units is $\frac{1}{N - 1}$ and so on. The probability of selecting a specified unit of the population at any given draw is equal to the probability of its being selected at the first draw.

Although this sampling method is not the easiest to organize, it is still the most commonly used. Simple random sampling protects against known as well unknown defects. It also protects against cycles and patterns, has low inherent risk, offers good representation of homogeneous lots, is highly reliable, and requires no special skills to operate.

Drawing a Simple Random Sample. Simple random samples can be drawn using tables of random numbers. Numerous random number tables are available for use. A table of random numbers is shown in Appendix Table 19 abstracted from the Rand Corporation's "A Million Random Digits with 100,000 Normal Deviates." For larger populations and repeated use of random numbers, consult the Rand Corporation tables. Random number tables are constructed in such a way that each of the digits 0, 1, 2, . . . , 9 appears with approximately the same frequency and independently of each other.

The method of selecting the random sample consists of the following steps:

First we identify the N units in the population to be sampled by numbering the N units from 1 to N. Next we select at random any page of the random number tables and, using any row, column, or diagonal, we select n random numbers. The units in the lot corresponding to the numbers selected constitute the random sample.

Example 12.1 In a creamery producing skim milk powder, the quality control manager wishes to select a random sample of n = 5 units from a lot containing the morning production run of 150 1-lb bags of skim milk powder.

Since the lot size (N = 150) has three digits, each assigned number is in the three-digit form, i.e., 001, 002, 003, . . . , 150. We assign numbers from 1 to 150 to each bag in the lot. Next we locate a starting point in the table of random numbers. If we are drawing only one sample, we may start at the beginning of the table. If, however, several samples are to be drawn, it is advisable to either take a pencil and with closed eyes select a point in the table at which to start each time or, alternatively, to start at the beginning of the table for the first sample, mark the point reached after the sample has been drawn, start from that point for the second sample, and so forth. (Note that the layout of numbers in groups of five within the table is simply for reading convenience.) For the present example, if we are working through the table by columns, the first five random numbers are

931, 282, 218, 220, 092

The numbers 931, 282, 218, and 220 should be neglected as being over 150 (the lot size). We proceed along the columns until we encounter other three-digit numbers between 001 and 150 which complete our sample of size five. The next numbers between 001 and 150 are

011, 066, 027, 149

The final five random numbers selected to make up the sample are

92, 11, 66, 27, 149

Hence, the units in the lot of 150 bags of skim milk powder corresponding to the numbers selected constitute the random sample.

If a random number selected from the table exceeds the lot size, we can (a) either neglect that number and proceed in the table to select another number between 001 and N, as was done in this example, or (b) divide the number by the lot size and take the remainder as the selected random number. For example, in example 12.1, we could divide 931 by 150 to get the remainder 31 as our random number. We may also draw the same three-digit numbers twice within the range prescribed by the lot size. To give each unit an equal chance of being included in the sample, it is preferable to take the number once and neglect it on repetition.

Estimation of Population Parameters. The information contained in the simple random sample helps draw inferences about the population parameters. Consider, for instance, a simple random sample $X_1, X_2, \ldots, X_n$ of size n drawn from a population of size N. The sample mean and variance are given by

$$\overline{X} = \frac{\sum_{i=1}^{n} X_i}{n}$$

and
$$S^2 = \frac{\sum_{i=1}^{n} (X_i - \overline{X})^2}{n - 1}$$

As shown, the sample mean $\overline{X}$ is an unbiased estimator of the population mean μ,

$$E\,(X) = \mu$$

The estimator $\hat{\mu}$ of the population mean μ is given by

$$\hat{\mu} = \overline{X} = \frac{\sum_{i=1}^{n} X_i}{n}$$

The estimator $\hat{T}$ of the population total T is given by

$$\hat{T} = N\overline{X} = N\,\frac{\sum_{i=1}^{n} X_i}{n}$$

The estimated variances of $\overline{X}$ and $\hat{T}$ are given by

$$\hat{V}(\overline{X}) = \frac{S^2}{n}\left(\frac{N - N}{N}\right)$$

and
$$\hat{V}(\hat{T}) = N^2\,\frac{S^2}{n}\left(\frac{N - n}{N}\right)$$

where $\frac{N - n}{N}$ is called the finite population correction (fpc). When n is small as compared to N, the fpc is close to unity. For all practical purposes the fpc is ignored if $n \leq \frac{N}{20}$. In that case $\hat{V}(\overline{X}) = \frac{S^2}{n}$, and $\hat{V}(\hat{T}) = N^2\,\frac{S^2}{n}$.

The population variance of $\overline{X}$ uses a slightly different fpc and is given by

$$V(\overline{X}) = \frac{\sigma^2}{n}\left(\frac{N - n}{N - 1}\right)$$

As explained in the introduction to this Chapter, the bounds on the error of estimation can be given as

$$e = 2\,\sigma_{\hat{\theta}} = 2\sqrt{V(\overline{X})}$$

$$= 2\sqrt{\frac{\sigma^2}{n} \cdot \frac{N - n}{N - 1}}$$

From here we can find the *sample size* required to estimate μ as

$$n = \frac{4N}{4 + \dfrac{(N - 1)e^2}{\sigma^2}}$$

In case n is small relative to N and the fpc is close to unity, the sample size required to estimate μ is given by

$$n = \frac{4\sigma^2}{e^2}$$

In the case of simple random sampling of attributes where we are interested in the estimation of a population proportion p using its corresponding sample estimator $\hat{p}$, we proceed as follows:

If a simple random sample of size n is drawn, the sample proportion $\hat{p}$ would be the fraction of the elements in the sample that possess the requisite characteristic. The properties of $\hat{p}$ are similar to $\overline{X}$, the sample mean. We can define the response measurement as follows:

Let $X_i = 0$ if the *i*th element sampled does not possess the specified characteristic and $X_i = 1$ if it does. Then $\hat{p}$ is the average of the 0 and 1 values from the sample and, therefore,

$$\hat{p} = \frac{\sum_{i=1}^{n} X_i}{n} = \overline{X}$$

Thus $\hat{p}$, the estimator of the population proportion p, is

$$\hat{p} = \overline{X} = \frac{\sum_{i=1}^{n} X_i}{n}$$

The estimated variance of $\hat{p}$ is given by

$$\hat{V}(\hat{p}) = \frac{\hat{p}\,\hat{q}}{n - 1}\left(\frac{N - n}{N}\right)$$

where $\hat{q} = 1 - \hat{p}$.
The corresponding formula for finding the sample size required to estimate p is given by

$$n = \frac{4N}{4 + \dfrac{e^2 (N - 1)}{pq}}$$

where $q = 1 - p$.

Example 12.2 A simple random sample of 10 cans of canned pears is selected from a lot containing 300 cans. The net weight in ounces of the contents in the cans are given here:
16.2, 17.0, 16.5, 16.0, 15.9, 18.1, 18.0, 16.1, 16.5, 17.6
Estimate μ, the average net weight, and place a bound on the error of estimation.

We have $n = 10$, $N = 300$. For the calculation of mean and variance, we have

$$\sum_{i=1}^{n} X_i = 167.4 \text{ and } \sum_{i=1}^{n} X_i^2 = 2825.33$$

The estimate of μ is given by

$$\overline{X} = \frac{\sum_{i=1}^{n} X_i}{n} = \frac{167.4}{10} = 16.74 \text{ oz}$$

To find a bound on the error of estimation we must calculate

$$S^2 = \frac{\sum_{i=1}^{n} X_i^2 - \dfrac{\left(\sum_{i=1}^{n} X_i\right)^2}{n}}{n - 1} = \frac{2825.33 - \dfrac{(167.4)^2}{10}}{9}$$

$$= 1.6005$$

Therefore, the bound on the error of estimation is given by

$$2\sqrt{\hat{v}(\bar{x})} = 2\sqrt{\frac{S^2}{n} \cdot \left(\frac{N - n}{N}\right)}$$

$$= 2\sqrt{\frac{1.6005}{10} \cdot \left(\frac{300 - 10}{300}\right)}$$

$$= 0.9952$$

The estimate of the mean net weight is 16.74 oz and the error of estimation is less than 0.9952 oz.

Example 12.3 If a product is 4% defective on the average, what sample size is required out of a lot of 400 so that there will be 95% confidence that the error in the estimate will not exceed 3%?

We have

$$p = 0.04, q = 1 - p = 0.96, N = 400,$$
$$\text{and } e = 0.03$$

Thus,

$$n = \frac{4N}{4 + \dfrac{e^2 (N - 1)}{pq}} = \frac{4 \times 400}{4 + \dfrac{(0.03)^2 \times 399}{0.04 \times 0.96}}$$

$$= 119.84 \simeq 120$$

Therefore, under the conditions specified, a sample of size 120 is required from a lot containing 400 items.

Systematic Sampling

Systematic sampling is sometimes considered a variation on simple random sampling. It is a frequently used method of sampling when a complete list of the sampling units is available. The method consists in selecting only the first unit at random, the remaining units being automatically selected according to some predetermined pattern involving regular spacing of units. Let us consider N sampling units numbered serially from 1 to N from which a sample size of n is to be drawn. We find an integer k, called the sampling interval, such that $k = \frac{N}{n}$, and then select randomly a number c between 1 and k. Then the required systematic sample is given by

$$c,\ c + k,\ c + 2k,\ \ldots,\ c + (n - 1)k$$

Strictly speaking, systematic sampling is partly probabilistic and partly nonprobabilistic. It is probabilistic in that the first unit is selected at random from the k units, and nonprobabilistic because the other units in the sample are fixed by the choice of the first unit and the sampling interval. An alternative is to use *circular systematic sampling* which consists in choosing the first unit from N items at random and then taking every *k*th unit in a circular manner until the whole list is exhausted. With systematic sampling the individuals have an equal chance of being selected, but their selection may not be independent of each other.

Systematic sampling has some advantages over simple random sampling in that it is much easier and quicker to operate, and provides better estimates if the sampling frame is arranged wholly at random. On the other hand, a systematic sample can be unrepresentative if there is a steady or cyclical trend in the population.

Example 12.4 A producer wishes to sell 500 boxes of apples and requires that the lot be inspected by a government inspector. The boxes of apples are to be taken out of the cold storage through the conveyor belt and loaded onto a truck. The inspector wishes to draw a sample of 25 boxes. Using systematic sampling, the inspector predetermines the following scheme:

$$\text{Lot size} \quad (N) = 500$$

$$\text{Sample size} \quad (n) = 25$$

$$\text{Sampling interval } (k) = \frac{N}{n} = \frac{500}{25} = 20$$

The inspector selects a random number c between 1 and 20; say it is 7. The boxes to be selected to make up a systematic sample of size 25 will be

$c, c + k, c + 2k, \ldots, c + 19k$

or $7, 7 + 20, 7 + 2 \times 20, \ldots, 7 + 19 \times 20$

or 7, 27, 47, 67, 87, 107, 127, 147, 167, 187
207, 227, 247, 267, 287, 307, 327, 347, 367, 387,
407, 427, 447, 467, 487

The inspector stands near the conveyor belt and selects the sample as the boxes move on the conveyor belt by first choosing box number 7 and then every 20th box after that until all 25 boxes comprising the sample size are chosen.

Estimation of Population Parameters. Unlike simple random sampling, the procedures for estimating the population parameters in the case of systematic sampling become more involved. For instance, an unbiased estimate of the variance of the mean of a systematic sample cannot be obtained using the data from only one systematic sample. However, with random populations whose elements are in random order, a systematic sample can be considered equivalent to a simple random sample, and we can use the same estimating procedure as used in the case of simple random sampling.

We denote by $\overline{X}_{sy}$ and S^2_{sy} the sample mean and variance, respectively, from a systematic sample of size n drawn from a lot of size N. As before, the estimators of the population mean μ and total T are given by

$$\hat{\mu} = \overline{X}_{sy} = \frac{\sum_{i=1}^{n} X_i}{n}$$

and

$$\hat{T} = N\overline{X}_{sy} = N\frac{\sum_{i=1}^{n} X_i}{n}$$

The estimated variances of $\overline{X}_{sy}$ and $\hat{T}$ are given by

$$\hat{V}(\overline{X}_{sy}) = \frac{S^2_{sy}}{n}\left(\frac{N - n}{N}\right)$$

and

$$\hat{V}(\hat{T}) = n^2\frac{S^2_{sy}}{n}\left(\frac{N - n}{N}\right)$$

The population variance of $\overline{X}_{sy}$ is given by

$$V(\overline{X}_{sy}) = \frac{\sigma^2}{n}[1 + (n - 1)\rho]$$

where ρ is the correlation coefficient between observations in the same systematic sample. If ρ is 0 and N is large, the population variance of the mean in a systematic sample $V(\overline{X}_{sy})$ is the same as in the case of a simple random sample $V(\overline{X})$. As we remarked before, an unbiased estimate of $V(\overline{X}_{sy})$ cannot be obtained from one systematic sample. However, if the units in the population are in random order, the elements of the systematic sample would be expected to be heterogeneous with ρ, that is, approximately equal to 0. The systematic sample and simple random sample would be equivalent in this case. The same holds true for an ordered population. In an ordered population, the units are ordered in magnitude in accordance with some criterion. A systematic sample drawn from this population will be heterogeneous with $\rho \leq 0$. This means that if N is large and $\rho \leq 0$, $V(\overline{X}_{sy})$ is less than or equal to $V(\overline{X})$; in this case, a systematic sample provides better information than can be obtained from a simple random sample. Finally, if the population is periodic, that is, it has a steady or cyclical trend, the elements of the systematic sample drawn from such a population are expected to be homogeneous and it would result in a larger variance than could be obtained through simple random sampling. One way of avoiding this problem is to select a random starting point several times throughout the systematic sample.

Example 12.5 In an experimental farm where 1000 tomato plants are grown, researchers wish to estimate both the average and the total yield of the crop. They select a sample of 7 plants using systematic sampling and record the yield per plant in grams as follows:

Plant number:	1	2	3	4	5	6	7
Yield (gm):	1280	1435	1512	1320	1295	1378	1442

Use these data to estimate μ, the average yield, and T, the total yield, and place a bound on the error of estimation.

From the data we calculate $\overline{X}_{sy}$ and S_{sy}^2 as follows:

$$\overline{X}_{sy} = 1380.2857, \; S_{sy}^2 = 7520.2370$$

An estimate of μ is given by

$$\overline{X}_{sy} = 1380.2857 \text{ gm}$$

and a bound on the error estimation is given by

$$2\sqrt{\hat{v}(\overline{X}_{sy})} = 2\sqrt{\frac{S_{sy}^2}{n} \cdot \frac{N-n}{N}}$$

$$= 2\sqrt{\frac{7520.2370}{7} \cdot \frac{1000-7}{1000}} = 65.3278$$

An estimate of T is given by

$$N\overline{X}_{sy} = 1000 \times 1380.2857 = 1380285.7$$

and a bound on the error of estimation is given by

$$2\sqrt{\hat{V}(N\overline{X}_{sy})} = 2\sqrt{N^2 \cdot \frac{S_{sy}}{N} \cdot \frac{N-n}{N}}$$

$$= 2\sqrt{1000^2 \cdot \frac{7520.2370}{7} \cdot \frac{1000-7}{1000}}$$

$$= 2065.7195$$

To summarize, we estimate the average yield to be 1380.29 gm and the total yield to be 1380285.7 gm or 1380.2857 kg.

Stratified Random Sampling

In stratified random sampling the units in the lot or population are divided into nonoverlapping groups or parts called *strata:* then a simple random sample is drawn from each stratum. The emphasis is on dividing the lot into subgroups so that the units within each stratum are more homogeneous than within the population as a whole. With reduced variability within each stratum, a stratified random sample produces better estimates than a corresponding simple random sample. Also, since each stratum has fewer units or covers a smaller geographic area than does the whole population, it is administratively more convenient to operate and is less costly than simple random sampling.

Estimation of Population Parameters. Consider a population consisting of N sampling units. It is divided into k relatively homogeneous strata of size $N_1, N_2, \ldots, N_k$, such that $N = \sum_{i=1}^{k} N_i$. The stratified sample is composed of k partial samples drawn at random from the corresponding strata. We denote, by $\overline{X}_i$, the sample mean for the simple random sample of size n_i, selected from stratum i, μ_i the population mean for stratum i, and T_i the population total for stratum i. The sample variances for the k strata are denoted by S_1^2, $S_2^2, \ldots, S_k^2$, where

$$S_i^2 = \frac{\sum_{i=1}^{n_i} (X_{ij} - \overline{X}_i)^2}{n_i - 1}, i = 1, 2, \ldots, k$$

These variances estimate the corresponding true stratum variances $\sigma_1^2, \sigma_2^2, \ldots, \sigma_k^2$.

The estimators of the population mean μ and population total T are given by

$$\overline{X}_{st} = \frac{1}{N}[N_1\overline{X}_1 + N_2\overline{X}_2 + \ldots + N_k\overline{X}_k] = \frac{1}{N}\sum_{i=1}^{k} N_i\overline{X}_i$$

and $$N\overline{X}_{st} = N_1\overline{X}_1 + N_2\overline{X}_2 + \ldots + N_k\overline{X}_k = \sum_{i=1}^{k} N_i\overline{X}_i$$

respectively, where $\overline{X}_{st}$ denotes the estimator of μ when stratified random sampling is used.

The estimated variances of $\overline{X}_{st}$ and $N\overline{X}_{st}$ are given by

$$\hat{V}(\overline{X}_{st}) = \frac{1}{N^2}\sum_{i=1}^{k} N_i^2 \left(\frac{N_i - n_i}{N_i}\right) \cdot \frac{S_i^2}{n_i}$$

and $$\hat{V}(N\overline{X}_{st}) = \sum_{i=1}^{k} N_i^2 \left(\frac{N_i - n_i}{N_i}\right) \frac{S_i^2}{n_i}$$

When estimating population proportion p, we proceed as follows: we denote by $\hat{p}_i$ the proportion of characteristics under study for stratum i. Then $\hat{p}_i$ is an unbiased estimator of p_i, the population proportion in stratum i. The estimator of the population proportion p is given by

$$\hat{p}_{st} = \frac{1}{N}[N_1\hat{p}_1 + N_2\hat{p}_2 + \ldots + N_k\hat{p}_k] = \frac{1}{N}\sum_{i=1}^{k} N_i\hat{p}_i$$

and the estimated variance of $\hat{p}_{st}$ is given by

$$\hat{V}(\hat{p}_{st}) = \frac{1}{N^2}\sum_{i=1}^{k} N_i^2 \left(\frac{N_i - n_i}{N_i}\right) \frac{\hat{p}_i\hat{q}_i}{n_i - 1}$$

Example 12.6 In a shell egg packing plant, the boxes of eggs are placed on pallets according to the grade size. Pallet 1 has 200 boxes of grade large eggs, pallet 2 has 350 boxes of grade medium eggs, and pallet 3 has 150 boxes of grade small eggs. An inspector, wishing to estimate the overall egg quality, uses a random number table and selects a simple random sample of 5 boxes from each pallet. Each box contains 4 trays of 25 eggs each, a total of 100 eggs per box. The results are given as follows:

Pallets	Number of boxes in the pallet (N_i)	Sample size (n_i) Number of boxes	Number of eggs	Percent defective eggs in the sample
1	200	5	500	4%
2	380	5	500	5%
3	150	5	500	6%

The estimate of the population proportion is given by

$$\hat{p}_{st} = \frac{1}{N}[N_1\hat{p}_1 + N_2\hat{p}_2 + N_3\hat{p}_3]$$

$$= \frac{1}{700}[200 \times 0.04 + 350 \times 0.05 + 150 \times 0.06]$$

$$= 0.05 = 5\%$$

The estimated variance of $\hat{p}$ is given by

$$\hat{V}(\hat{p}_{st}) = \frac{1}{N^2}\sum_{i=1}^{k} N_i^2\left(\frac{N_i - n_i}{N_i}\right)\frac{\hat{p}_i\,\hat{q}_i}{n_i - 1}$$

$$= \frac{1}{(700)^2}\left[(200)^2\left(\frac{200 - 5}{200}\right)\left(\frac{0.04 \times 0.96}{5 - 1}\right)\right.$$

$$+ (380)^2\left(\frac{380 - 5}{380}\right)\left(\frac{0.05 \times 0.95}{5 - 1}\right)$$

$$\left.+ (150)^2\left(\frac{150 - 5}{150}\right)\left(\frac{0.06 \times 0.94}{5 - 1}\right)\right]$$

$$= 0.0048$$

The bound on the error of estimation is given by

$$2\sqrt{\hat{V}(\hat{p}_{st})} = 2\sqrt{0.0048} = 0.1392$$

The estimate of the average percentage of defective shell eggs is 5%, and we are confident that the bound on the error of estimation is less than 0.1392.

Exercises

1. Of the 100 milk samples received at a laboratory 6 samples were selected by the simple random sampling method and analyzed for fat content. The following fat percentages were obtained:

   3.52, 3.56, 3.65, 3.60, 3.68, 3.59

   Estimate the average fat percentage of milk and place a bound on the error of estimation.

2. The infection rate of a certain disease in animals is known to be 35%. What sample is required to estimate the true infection rate with a high level of confidence where it is desired that the error in the estimate will not exceed 2%?

3. In an experimental orchard containing 750 apple trees, researchers wish to estimate the total yield from the orchard. Using systematic sampling, they select a sample of 20 trees giving a sample mean and variance of $\overline{X}_{sy} = 4.1$ bushels and $S^2_{sy} = 0.62$ bushels. Estimate the total yield T and place a bound on the error of estimation.

4. Ten 50-lb bags of skim milk powder were randomly selected from a lot containing 300 bags. A small sample of the milk powder was taken from each of the 10 bags for laboratory analysis. The results for percent moisture for the 10 samples are as follows:

   3.1, 2.7, 3.9, 4.2, 3.4, 3.2, 4.1, 2.9, 3.3, 3.5

   Using these data, estimate the average percentage of moisture in skim milk powder, and place a bound on the error of estimation.

5. A food company wants to introduce a new kind of instant soup. It selects 20 households in 4 areas or strata and distributes free samples of the product. After two weeks, the company representative surveys the number of households that bought that type of soup after having tried the free sample. The data are presented here:

Stratum	Sample Selected (n_i)	Number that responded by purchasing the product
1	20	12
2	20	8
3	20	10
4	20	7

   Use these data to estimate the true proportion and place a bound on the error of estimation.

6. A random sample of 10 fields of oats was selected and the number of poppy plants found in each of the 10 fields was recorded as follows:

   85, 70, 15, 75, 500, 20, 45, 250, 40, 36

   Estimate the population mean and total and place a bound on the errors of estimation.

13.

Statistical Quality Control

An investigator often has to deal with continuous processes that require control of error and variability. Consider, for example, a food scientist working in a food-producing company. The research emphasis is directed toward producing a high-quality product and achieving a state of control of that high-quality product over time. To realize this goal, the scientist has to develop a sampling inspection system for checking the quality and devise methods of process control. Data resulting from quality control programs have to be analyzed through suitable statistical techniques. In this sense of the term, statistical quality control (SQC) techniques become an inseparable part of basic applied statistical methods. It would be impossible to discuss all the methods of SQC in a single chapter, but the material presented here will acquaint the reader with the basic concepts.

Statistical quality control refers to the application of statistical methods to the problems of producing, controlling, and maintaining quality of product, process, or operation. The main objective of any production/manufacturing process is to produce a high-quality product that meets specifications and also is economical, reliable, and satisfactory. To achieve this objective, a food processing company has to establish a total quality control program that applies to all aspects of quality related to the product. These include quality control, quality assurance, and quality reliability. In this chapter we limit our discussion to two essential functions of total quality control: acceptance sampling and process control. If we look at the total quality process, we realize that once the product is fairly quality controlled, there are only two functions to be performed: sampling inspection or monitoring of the product, and implementation of methods of assessing and achieving a state of control of the product or process.

13.1 Acceptance Sampling

Acceptance sampling refers to the process of accepting or rejecting a lot by inspecting a sample selected in accordance with a predetermined sampling plan. A sampling plan specifies the amount of sampling to be done, the acceptance/rejection criteria, and the associated probabilities of acceptance. Sampling plans are based on several quality characteristics and the choice of a particular type of sampling plan depends on the nature of the product and the purpose of inspection. Selecting an adequate and suitable sampling plan, though important, is not always an easy task, because the selection is dependent on a number of different factors such as ease of administration of the sampling plan, protection afforded, amount of inspection required, cost of inspection, and the power of a plan to discriminate between a good and a bad lot. Before we discuss the methods of selecting the sampling plans, some important terms need to be defined.

13.1.1 Methods of Inspection

Measurement and evaluation of quality characteristics can be done by two methods, namely attributes and variables.

An *attribute* is a characteristic or property that is appraised in terms of whether it does or does not exist with respect to a given requirement. With attribute inspection, the units of product are classified as "defective" or "nondefective," "within tolerance" or "out-of-tolerance," or "correct" or "incorrect."

A *variable* is a characteristic or property that is appraised in terms of scalar values on a continuous scale. In variable inspection one measures quality characteristics such as weight, chemical purity, and temperature readings on a continuous numerical scale.

Attribute sampling plans are more commonly used methods of acceptance sampling than variable sampling plans because they are mathematically simple, economical, and administratively convenient. In this chapter we shall restrict our discussion to the derivation and usage of attribute sampling plans.

13.1.2 Characteristics of a Sampling Plan

Every sampling plan is characterized by a *sample size* (n), an *acceptance number* (c), and the *probability of acceptance* (P_a). The sample size is the number of items selected for inspection. The acceptance number is the largest number of defectives (or defects) in the sample that will permit the product to be accepted. The probability of acceptance of a sampling plan is the percentage of samples out of a long series of samples that will cause the product to be accepted. A complete plotting of the probability of acceptance for products at all possible levels of percent defective is known as an *operating characteristic (OC) curve.* One such OC curve is shown in figure 13.1. It can be seen from the figure that the probability of acceptance for product 2% defective is 0.95. This means that 95% of the product will be accepted.

13.1.3 Parameters Affecting Sampling Plans

Operating characteristic curves are a graphical means for showing the relationship between quality of lots submitted for sampling inspection and the probability of their acceptance. The steepness of the slope and shape of an OC curve are dependent on the sample size (n) and the acceptance number (c). The steepness of the OC curve indicates the power of the sampling plan to discriminate between good and bad quality. An increase in sample size results in a steepening of the OC curve, as indicated in figure 13.1.

The effect of changing the acceptance number is illustrated in figure 13.2. The larger the acceptance number, the poorer the

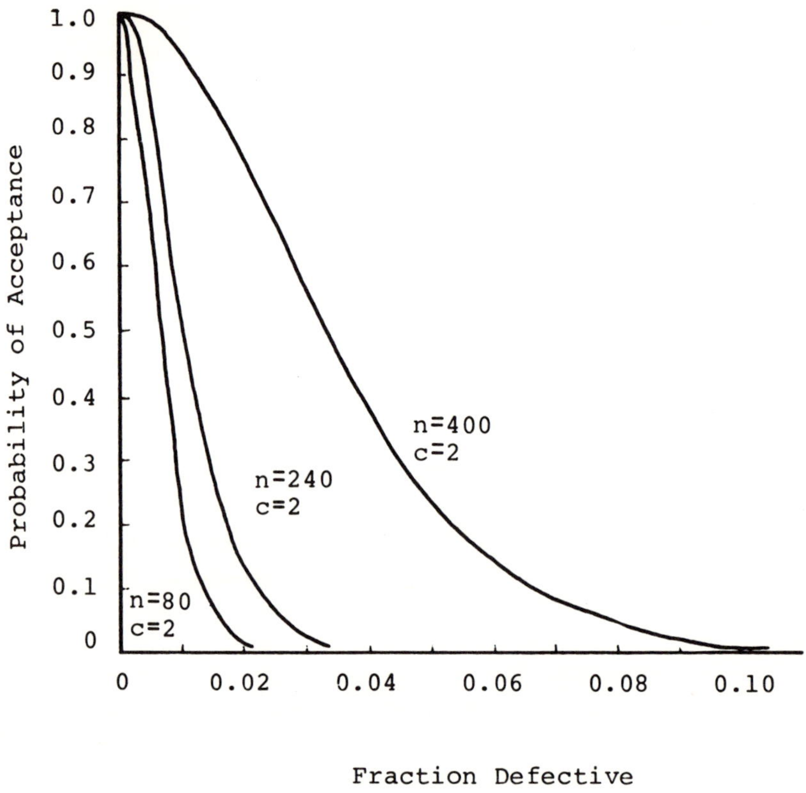

Fig. 13.1 *OC curves for samples of different size.*

quality that will be passed by the sampling plan. The effect of increasing the acceptance number is to shift the location of the entire OC curve to the right. To obtain a sampling plan that closely satisfies the quality requirements, the sample size and acceptance number have to be simultaneously adjusted.

The effect of the lot size (N) on a sampling plan is not too pronounced. The relationship between the lot size (N) and sample size (n) for a specific method of selecting the sample was discussed in the previous chapter.

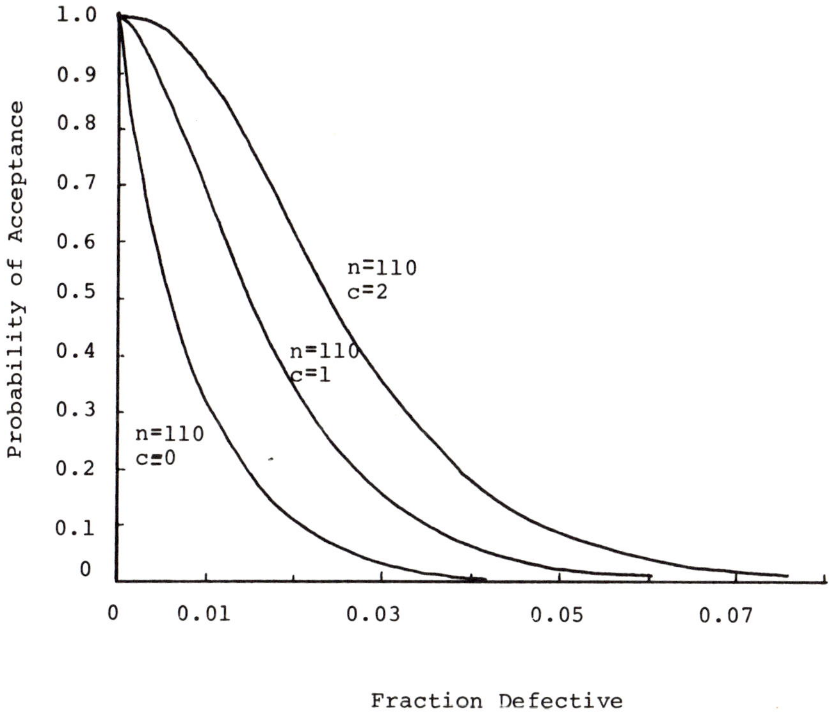

Fig. 13.2 *OC curves for different acceptance numbers.*

13.1.4 Construction of an OC Curve

Calculations of acceptance probabilities are done by using the hypergeometric, binomial, or Poisson distributions. We shall make use of the Poisson approximation to binomial for the derivation of these probabilities. Appendix Table 20 gives the summation of terms of Poisson's exponential binomial limit.

Example 13.1 Suppose we wish to calculate the probabilities of acceptance and draw the OC curve when

$$n = 240, \qquad \text{and} \qquad c = 2$$

Set up the table headings as in table 13.1:

Table 13.1 *Preparation for calculating acceptance probabilities in example 13.1.*

n	np′	p′	P_a
240			0.99
240			0.95
240			0.75
240			0.50
240			0.25
240			0.10
240			0.02
240			0.01

where n = sample size = 240
p′ = fraction defective
np′ = number defective,
and P_a = probability of acceptance = suitably chosen coordinates to facilitate construction of the OC curve

Now (using Appendix Table 20), find the np′ values for c = 2 corresponding to the P_a values closest to the P_a values chosen in table 13.1. For example, searching through the table, we find np′ = 0.44 for P_a = 0.990 under the column for c = 2. Similarly, np′ = 1.7 for P_a = 0.757 which is the closest value to P_a = 0.75 set in table 13.1. The same procedure is followed for each of the other P_a values and we obtain table 13.2:

Table 13.2 *Acceptance probabilities for example 13.1.*

n	np′	p′	P_a
240	0.44	0.0018	0.990
240	0.82	0.0042	0.950
240	1.70	0.0071	0.757
240	2.70	0.0113	0.494
240	3.90	0.0163	0.253
240	5.30	0.0221	0.102
240	7.50	0.0313	0.020
240	8.40	0.0350	0.010

The OC curve for table 13.2 is plotted in figure 13.2.

13.1.5 Types of Lot Sampling Plans

There are three important types of lot sampling plans: single, double, and multiple. The choice of which type to use depends on many factors such as quality history, quality requirements, economy, and the particular sampling inspection situation.

Single Sampling

In a single sampling plan a single sample of n items is selected at random from the lot. The decision concerning the acceptability of the lot is made on the basis of results obtained from the sample. If the number of defectives found in the sample is less than or equal to the acceptance number c, the lot is accepted. If the number of defectives is equal to or greater than the rejection number r, the lot is rejected.

Double Sampling

In a double sampling plan a first sample of n_1 units is selected at random from the lot and inspected. If the number of defectives is less than or equal to the first acceptance number c_1, the lot is accepted. If the number of defectives is equal to or greater than the first rejection number r_1, the lot is rejected. If no decision can be made from the first sample because the number of defectives is greater than c_1 but less than r_1, a second sample of n_2 units is selected at random from the lot and inspected. If the cumulative number of defectives from the first and second sample is less than or equal to the second acceptance number c_2, the lot is accepted. And if the cumulative number of defectives is equal to or greater than the second rejection number r_2, the lot is rejected.

The average number of items inspected with double sampling is generally less than that inspected with single sampling. Some inspectors consider double sampling equivalent to giving another chance to the product. Despite a smaller sampling rate, double sampling is less frequently used as compared to single sampling.

Multiple Sampling

The procedure in multiple sampling is similar to that in double sampling except that the number of successive samples required to reach a decision to accept or reject the lot may be more than two. The number of steps required to reach a firm decision depends on the cumulative number of defectives found in the samples taken progressively. There are acceptance/rejection criteria for each step—accept at any step where the cumulative defectives are equal to or less than the acceptance number, and reject where the defectives equal or exceed the rejection number for that step. If the number of defectives is between the accept/reject figures, we take another sample. All multiple sampling is terminated after a specified number of steps by arranging the acceptance and rejection figures to be consecutive at the last step, thus forcing a decision to accept or reject the lot. The size of the cumulative sample at the last step is larger than the equivalent in single and double plans.

13.1.6 Risks and Quality Indices

Regardless of the inspection plan used (sampling or 100% inspection), there is always a risk that a small percentage of defective units will be passed. Two types of risks are involved in sampling: risk of rejecting a good lot (Producer's Risk or α (alpha) risk), and risk of accepting a bad lot (Consumer's Risk or β (beta) risk).

The OC curve of a sampling plan quantifies these risks and makes it possible to state these risks numerically and describe the quantities of product that can be expected to be accepted if the quality standard is met, and the quantity rejected if the standard is not met. Ideally, we would like to have a sampling plan as illustrated in figure 13.3 when it is desired to accept all lots 3.0% defective or less and reject all lots having a quality level greater than 3.0% defective. All lots less than 3.0% defective will be accepted 100% of the time and all lots greater than 3.0% defective will be rejected 100% of the time. In practice, no sampling plan can discriminate perfectly; there is always the risk of rejecting good lots and accepting bad lots. We always start with an OC curve (figure 13.3) for an actual sampling plan and try to make the slope of the curve steeper by improving the quality of incoming lots.

Associated with the producer's risk is the quantity called *Acceptable quality level* (AQL), defined as the maximum percent defective (or maximum number of defects per 100 units) that, for the purpose of sampling inspection, can be considered as a process average. Since the acceptance/rejection of a lot is based on the inspection of a sample and involves the inherent risks associated with it, the producer is generally concerned about the probability of rejecting lots that meet the specifications and would not want a high

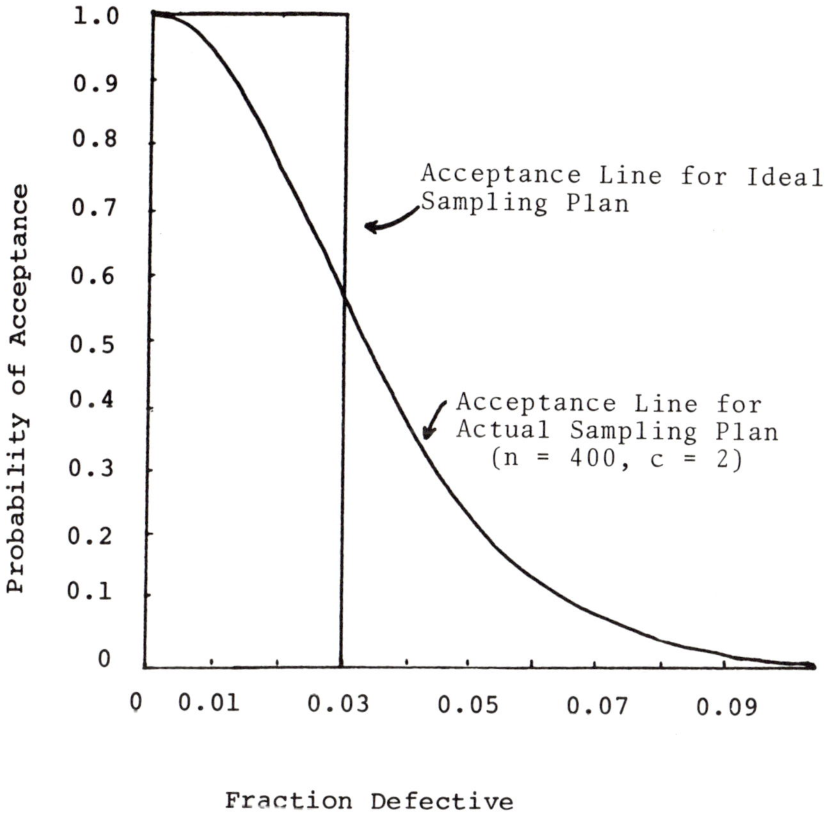

Fig. 13.3 *OC curves for ideal and actual sampling plans.*

quality lot to be rejected more than 5% of the time. Thus, the producer and the consumer agree on an AQL value and the associated α risk, consequently, a lot which is of AQL quality is accepted with a probability of $(1 - \alpha)$ or P_v. This gives us a point (AQL, P_v) on the OC curve.

The consumer, on the other hand, also sets a certain quality level, P_t for the incoming lot or process, called *lot tolerance percent defective (LTPD)*. This is the worst quality consumers would be willing to accept. It would be rarely accepted, or accepted with a probability of β or P_c, say 10%. Thus we have another point (P_t, P_c) on the OC curve.

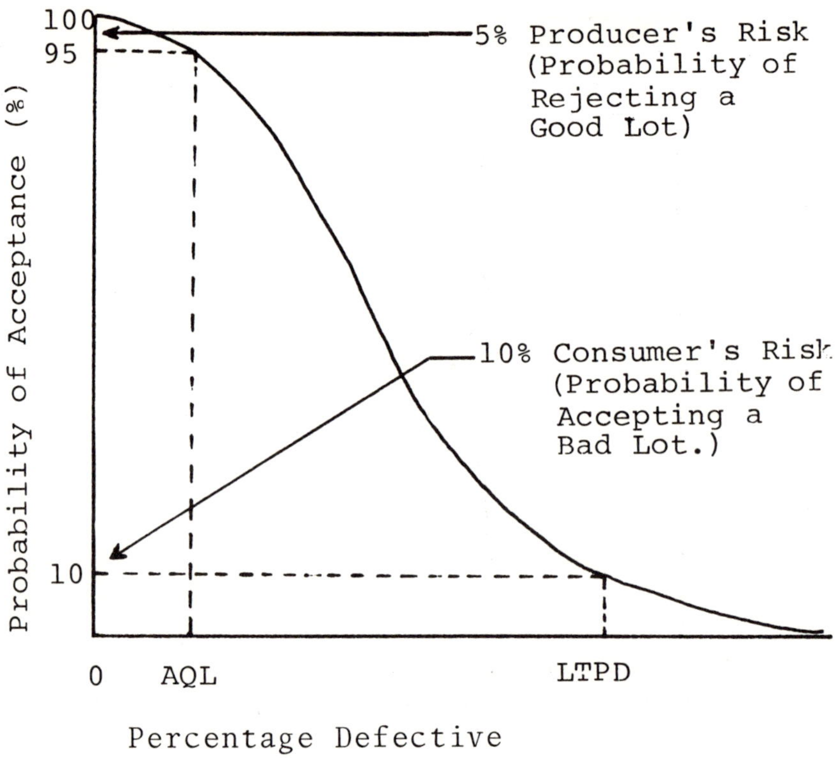

Fig. 13.4 *OC curve for attributes sampling.*

These two points (AQL, P_v) and (P_t, P_c) determine the shape of the OC curve for a sampling plan and are illustrated in figure 13.4. The OC curve that comes closest to passing through these two points, giving n and c, is the most suitable sampling plan for that situation. Because of the integral nature of these criteria, an exact solution to the desired specifications set by producer and consumer may not be realized. It is often necessary to satisfy one party and almost satisfy the other in selecting a sampling plan.

13.1.7 Choosing a Sampling Plan

A sampling plan can be derived using Appendix Table 20 and the two points (AQL, P_v) and (P_t, P_c). Once the values of these four characteristics are established, the OC curve is manipulated around these points until an optimum sampling plan results. This procedure is explained in the next section by considering an example.

Fortunately, we do not have to go through the rigor of deriving a sampling plan with the two points (AQL, P_v) and (P_t, P_c) every time. There are ready-made sampling tables available that can be used. The two most commonly used sampling tables are the U.S. Department of Defense *Military Standard 105D* (Mil Std 105D), based on AQLs, and *Dodge and Romig tables* based on LTPDs.

The AQL sampling plans are designed to protect the producer, whereas the LTPD sampling plans are designed to protect the consumer. Other types of sampling plans designed to protect the consumer and given in Dodge and Romig tables are known as average outgoing quality limit (AOQL) sampling plans. For a full description of LTPD and AOQL sampling plans, consult the Dodge and Romig tables cited in the bibliography.

In this text, we restrict our discussion to the use of Military Standard 105D (AQL sampling plans) as they are the most widely accepted methods of acceptance sampling.

Example 13.2 *Derivation of a single sampling plan:* Suppose a food processing company wishes to establish a single sampling plan for sampling inspection of cans of mixed vegetables. The producer agrees with the buyer (consumer) on an AQL value of 2% defective with $\alpha = 0.05$ or acceptance probability of 0.95. The buyer also specifies a LTPD value of 8% with $\beta = 0.10$ or acceptance probability of 0.10. This means that the producer wants to be 95% sure that the lot will be accepted if it meets the AQL specification of 2%, and the buyer wants to ascertain that the incoming lot will be accepted only 10% of the time if it is 8% or more defective. Thus, we must establish a value of n and c that will meet these specifications.

Using Appendix Table 20, the values of $np'_{0.95}$ and $np'_{0.10}$ are obtained for various values of c and the ratios $np'_{0.10}/np'_{0.95}$ are computed. The calculations are shown in the following table 13.3.

Next calculate the ratio

$$\frac{nPt_{0.10}}{nAQL_{0.95}} = \frac{0.08}{0.02} = 4.0$$

This value, when compared with the corresponding ratios in table

Table 13.3 *Acceptance calculations for example 13.2.*

c	$np'_{0.95}$	$np'_{0.10}$	$np'_{0.10}/np'_{0.95}$
0	0.05	2.3	45.10
1	0.35	3.8	10.96
2	0.82	5.3	6.50
3	1.37	6.7	4.89
4	1.97	8.0	4.06
5	2.60	9.3	3.55
6	3.30	10.5	3.21
7	4.00	11.8	2.96
8	4.70	13.0	2.77

13.3, lies between c of 4 and 5. Now consider four sets of probabilities resulting from combining each of c with LTPD and AQL values to calculate the corresponding n values. Calculations for the four sets of values are as follows:

(a) When c = 4, $P_c = 0.95$

$$n = \frac{nP'_{0.95}}{AQL} = \frac{1.97}{0.02} = 98$$

$$np' = n \times P_t = 98 \times 0.08 = 7.84$$

Thus, for np' = 7.84, and c = 4, $P_c = 0.112$ from Appendix Table 20.

(b) When c = 4, $P_c = 0.10$

$$n = \frac{nP'_{0.10}}{P_t} = \frac{8.0}{0.08} = 100$$

$$np' = n \times AQL = 100 \times 0.02 = 2.0$$

Thus, for np' = 2.00 and c = 4, $P_v = 0.947$ from Table 20.

(c) When c = 5, $P_v = 0.95$

$$n = \frac{np'_{0.95}}{AQL} = \frac{2.60}{0.02} = 130$$

$$np' = n \times P_t = 130 \times 0.08 = 10.4$$

Thus, for $np' = 10.4$ and $c = 5$, $P_c = 0.053$ from Table 20.

(d) When $c = 5$, $P_c = 0.10$

$$n = \frac{np'_{0.10}}{P_t} = \frac{9.5}{0.08} = 116$$

$$np' = n \times AQL = 116 \times 0.02 = 2.32$$

Thus, for $np' = 2.32$ and $c = 5$, $P_v = 0.97$ from Table 20.

The next step is to analyze the four sampling plans and choose the one that fits the given specifications most closely. In doing so, we can see that plan (a) gives the required protection for the producer, but increases the probability of accepting lots of quality equal to P_t. Plan (b) gives the required protection for the consumer and gives almost exact probability of accepting good lots at the 95% level. Plan (c) gives required protection to the producer but reduces the risk of accepting poor lots. Plan (d) gives the required protection to the consumer but increases the probability of accepting good lots.

Since plan (b) gives assurance at almost exact specifications, this is the plan of choice. If a sample of 100 items is selected and inspected, the lot is accepted if 4 or less items are found defective. If the number of defectives is greater than 4, the lot is rejected. The OC curve for the plan of choice (n, c = 100, 4) is shown in figure 13.5.

13.1.8 MIL STD 105D

The United States Department of Defense Mil Std 105D, entitled "Sampling Procedures and Tables for Inspection by Attributes" is published in two other forms: "Canadian Government Specification Board Standard CGSB 105-GP-1," and "ABC Standard Sampling Procedures and Tables for Inspection by Attributes." It represents the international standard agreed upon by the United States, Britain, and Canada. All three standards are identical with the exception of their titles. We present only the essential feature of this

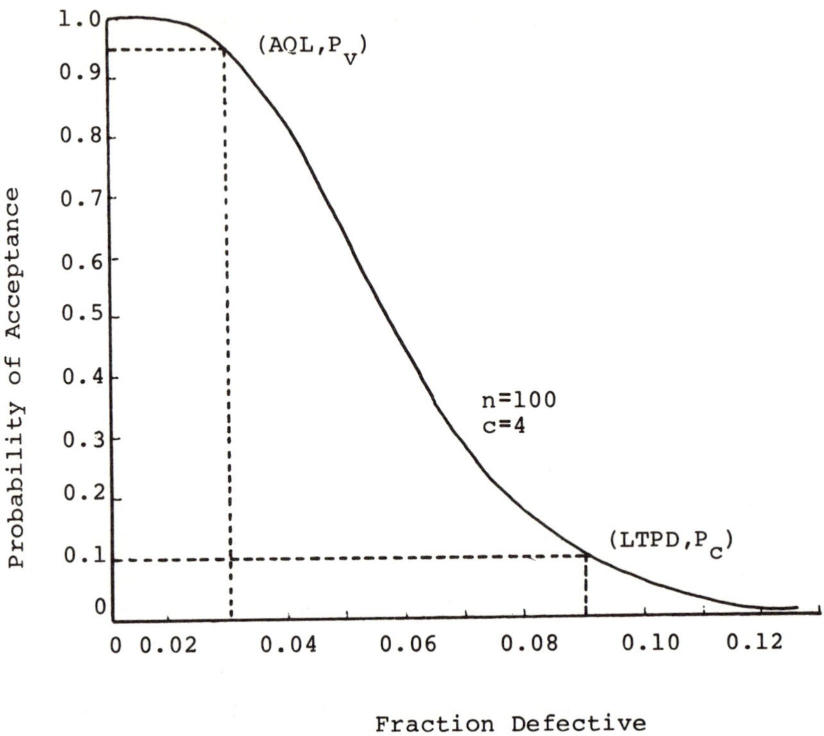

Fig. 13.5 *OC curve for single sampling plan.*

standard here. The reader is advised to consult the full text of the standard for a complete set of sampling tables.

Mil Std 105D provides sampling inspection plans by attributes based on AQLs designed to be applied to lots emerging from long production runs of many units of products. Information is also provided to allow for easy extraction of the protection afforded by these plans in terms of LTPD and AOQL.

Three types of sampling plans are provided in this standard: single, double, and multiple. The size of sample required for inspection is dependent on four factors: inspection level, lot size, type of sampling plan, and AQL.

Seven inspection levels, S-1, S-2, S-3, S-4, I, II, and III, are provided for varying degrees of discrimination and each level provides different sample sizes for a given lot size. In the order given above, sample size (therefore discrimination) increases from a minimum at level S-1 to a maximum at level III. Levels S-1 to S-4 are considered special levels, the use of which is limited to situations where it is imperative that only small sizes be used, such as destructive testing of expensive units of product. Inspection level II is considered the normal level and is to be used at the commencement of inspection activities unless otherwise specified.

The standard also provides three levels of inspection in terms of the severity of inspection: Normal, tightened, and reduced. Normal inspection is used to start. Then, if the quality is shown to be poor, the inspector is directed to be more severe with his inspection and uses the tightened inspection level. If the quality is shown to be consistently high, reduced inspection is indicated. Guidelines for switching rules between normal, tightened, and reduced inspections are provided.

A letter code system is used for determining sample size. This is given in Table I of the Mil Std 105D and has been produced in Appendix Table 21.1 in this book. The letter assigned to a given sample is dependent on the inspection level and lot size. In the table, varying blocks of lot sizes are listed vertically and inspection levels listed horizontally. Any specific lot size can be associated with a corresponding block listed in the table. Suppose we have a lot of size 350 and we wish to use general inspection level II. In Appendix Table 21.1, lot size from 281 to 500 is applicable to the lot size of 350; reading horizontally to the column below inspection level II, we find the letter H. This becomes the sample code letter for this specific inspection activity.

The next step is deciding the type of sampling plan and the AQL to be used. The standard provides single, double, and multiple sampling plans for normal, tightened, and reduced inspection levels. The AQLs progress in steps from a minimum of 0.010 to a maximum of 1000. AQLs from 0.010 up to 10 may be expressed in either percent defective or defects per 100 units; AQLs greater than 10 are expressed in defects per 100 units only.

Consider, for example, a government inspector wishing to establish a sampling plan for the inspection of heads of lettuce. The lot size is 250 heads, the AQL is specified at 4% defective, and it is desirable to use normal inspection level II and a single sampling plan.

In Appendix Table 21.1, the sample size code letter for a lot size 250 and general inspection level II is given as G. Turning to Appendix Table 21(b) (reproduced from Table IIA of Mil Std 105D), we find that the sample size for the code letter G is n = 32 and the acceptance and rejection numbers for this sample size for an AQL of 4% are (3,4). This sampling plan dictates that a sample of 32 items be selected and inspected out of a lot size of 250. If the number of defective items found in the sample is less than or equal to 3, the lot is accepted. If the number of defective items is equal to or greater than 4, the lot is rejected.

In this text we are only giving data for single sampling plans for normal inspection. For other sampling tables, the reader should consult the Standard. The procedure for double and multiple sampling plans is similar to that for single sampling plans.

The OC curves for the sampling plans are also contained in the Standard. There is a family of curves for each sample code letter. Each curve in the family is related to an AQL and a curve can be chosen that is related to the specific inspection of interest.

13.2 Control Charts

The second major task in a quality control program after sampling inspection is process control. This requires studying the patterns of variation in data resulting from a continuous production process. There are two types of variations in process quality, random and nonrandom. Random variations, also known as natural tolerances of the process, are due to chance causes and cannot be altogether eliminated. Nonrandom variations or assignable causes can be identified and eliminated. Examples of nonrandom causes are improper performance of equipment, maladjustment of machinery, new workers, lack of uniformity in material or workmanship, and deterioration.

In 1924 Dr. Walter A. Shewhart of Bell Telephone Laboratories proposed the method of control charts for studying unnatural patterns of variations and dealing with related problems of controlling quality during production. The control chart is a graphic method of presenting and evaluating ongoing product quality. A graph that reflects the state of product quality allows us to monitor the process and take action when the process is out of control with reference to preset specifications; this helps bring about a state of control in a process. With control, the performance of the product is predictable; it is governed by random causes alone, the inspection costs are low, wastage is minimum, and reliability is high. Control charts are also used in establishing a new specification, revising an existing one, and assessing whether a tolerance is a true reflection of the quality actually being produced or capable of being produced.

13.2.1 Nature of Control Charts

The control chart is a graphic representation of observations or means of the group of observations in the order of production. The selection of subgroups is important; it should be chosen so that the variations within a subgroup may be attributed to chance and the variability between the subgroups represents the fluctuations due to assignable causes. The choice of sample size and the frequency of sampling depend on the particular situation under consideration. However, it is suggested that 15 to 25 samples of size 4 to 6 provide good estimates.

The general configuration of a control chart is shown in figure 13.6. It consists of three horizontal lines: (i) a central line indicating an average value or a desired quality level, (ii) an upper control limit, and (iii) a lower control limit. Shewhart has proposed the control limits at 3 σ distance on each side of the central line.

It is assumed throughout the discussion of control charts that when the variates are in a state of statistical control, they are often normally distributed. With the help of Appendix Table 3 (areas under the normal curve) and the normal theory discussed in chapter 5, we can calculate the probability of a sample mean $\overline{X}$ falling between the two limits.

13.2.2 Types of Control Charts

Like the sampling inspection methods discussed in section 13.1.1, control charts are of two types: variables and attribute. We con-

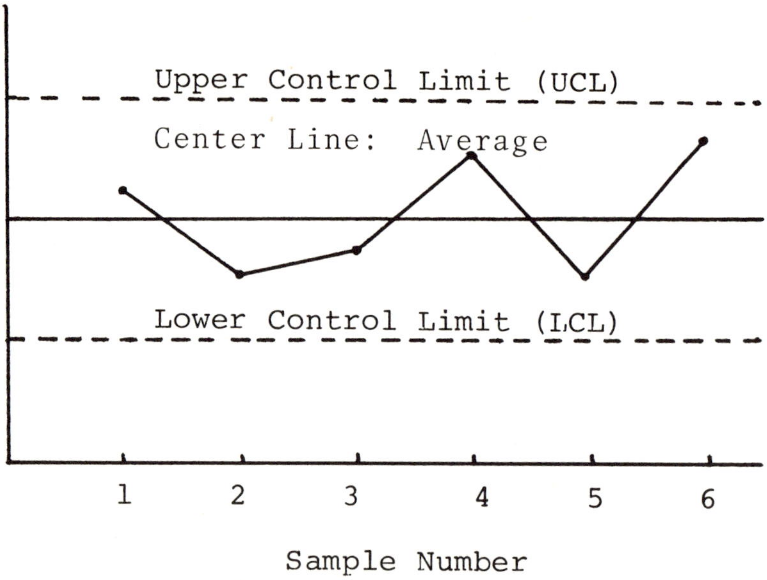

Fig. 13.6 *Outline of a control chart.*

sider the following control charts here:

1. Variables control charts: mean chart ($\overline{X}$-chart), and range-chart (R-chart).
2. Attribute control charts: fraction defective chart (p-chart).

$\overline{X}$-Charts

The means of samples taken at regular intervals from a production process are plotted on the $\overline{X}$-chart. The $\overline{X}$-chart shows where the process is centered and is an indicator of the stability of the process. The $\overline{X}$-chart is used to study past data or to control existing quality. When studying past data, the $\overline{X}$-chart is set as follows:

$$\text{center line} = \overline{\overline{X}} = \text{the grand mean of all the data}$$

$$\text{upper control limit (UCL)} = \overline{\overline{X}} + 3\,\frac{S}{\sqrt{n}}$$

$$\text{lower control limit (LCL)} = \overline{\overline{X}} - \frac{3S}{\sqrt{n}}$$

where S is an estimate of σ, the population standard deviation obtained from the past data.

The value of S can be determined from the mean range $(\overline{R})$ as

$S = \frac{\overline{R}}{d_2}$, where d_2 is the value obtained from Appendix Table 22 corresponding to a sample size of n. In setting up an $\overline{X}$-chart for controlling existing quality, we can use the mean and standard deviation for past data or some other value set by management. When past data are not available, a preliminary run of ongoing quality is made, and the data obtained provide the mean and the standard deviation.

R-Charts

While the $\overline{X}$-chart reveals undesirable variations between the average of samples, the R-chart reveals undesirable variation within samples. The R-chart is an indicator of the magnitude of the variability of the process being studied. It is a measure of consistency or uniformity of the process. The R-chart is set as follows:

center line = $\overline{R}$ = the mean range, i.e., mean of all the sample ranges

$$UCL = \overline{R} + 3S_R$$

$$LCL = \overline{R} - 3S_R$$

where S_R is an estimate of the standard deviation of the range and is calculated as

$$S_R = d_3 S$$

where S is an estimate of the population standard deviation and values of d_3 are given in Appendix Table 22.

Example 13.3 A food processing company produces 19.5 fl oz cans of tomato juice and wishes to study the state of process quality. A preliminary study is done to determine the existing mean and standard deviation. Ten samples, each consisting of 6 cans, are taken and the values of means and ranges are calculated, as given in table 13.4.

Table 13.4 *Preliminary means and ranges of example 13.3.*

Sample Number	1	2	3	4	5	6	7	8	9	10	Mean
Mean	20.38	19.56	19.69	20.40	20.41	20.11	18.85	19.60	19.45	20.01	$\bar{\bar{X}} = 19.86$
Range	1.1	2.1	1.7	1.3	0.8	1.2	2.0	0.9	2.3	1.0	$\bar{R} = 1.44$

For the $\bar{X}$-chart,

$$S = \frac{\bar{R}}{d_2} = \frac{1.44}{2.534} = 0.57$$

The value of $\bar{R}$ comes from table 13.4 and the value of d_2 is taken from Appendix Table 22 for n = 6. For the R-chart,

$$S_R = d_3 S$$
$$= 0.8480 \times 0.57 = 0.48$$

The value of S = 0.57 is taken from the previous calculation and the value of d_3 is taken from Appendix Table 22 for n = 6.

The control limits can now be calculated as follows:

1. $\bar{X}$-chart:

$$\text{center line} = \bar{\bar{X}} = 19.86$$

$$\text{UCL} = \bar{\bar{X}} + \frac{3S}{\sqrt{n}} = 19.86 + \frac{3 \times 0.57}{\sqrt{6}}$$
$$= 20.56$$

$$\text{LCL} = \bar{\bar{X}} - \frac{3S}{\sqrt{n}} = 19.86 - \frac{3 \times 0.57}{\sqrt{6}}$$
$$= 19.16$$

2. R-chart:

$$\text{center line} = \bar{R} = 1.44$$

$$\text{UCL} = \bar{R} + 3S_R$$
$$= 1.44 + 3 \times 0.48 = 2.88$$

$$\text{LCL} = \bar{R} - 3S_R$$
$$= 1.44 - 3 \times 0.48 = 0$$

The quality control supervisor now takes a sample of 6 cans every 1-hr shift for 8 shifts in a day's production. The data and the calculations of means and ranges are given in table 13.5.

Table 13.5 *Means and ranges from full sample in example 13.4.*

Shift number	Number in sample 1	2	3	4	5	6	Mean	Range
1	20.0	20.2	20.4	19.3	20.5	20.3	20.12	1.2
2	19.1	20.0	18.6	20.9	19.9	18.7	19.53	2.3
3	19.7	20.2	20.2	19.4	19.9	18.9	19.72	1.3
4	20.4	19.0	19.7	21.0	19.7	20.3	20.02	2.0
5	20.0	20.4	19.3	20.2	20.3	20.5	20.12	1.2
6	20.0	20.6	20.8	20.7	20.5	20.1	20.45	0.8
7	21.0	20.8	19.9	20.5	20.6	20.8	20.60	1.1
8	20.9	20.8	20.7	21.1	20.9	20.6	20.83	0.5

Figure 13.7 shows the $\overline{X}$-chart with control lines and the sample mean values from table 13.5. Figure 13.8 shows the R-chart with control lines and the sample range values from table 13.5.

As shown in the $\overline{X}$-chart, the means fluctuated around the control chart mean most of the day. However, toward the end of the day, the filler seemed to malfunction and a corrective action was required. The values in the R-chart seem to be distributed randomly around $\overline{R}$ line and well within the control limits.

Remarks on $\overline{X}$- and R-Charts

A chart tells at a glance if operation variations are significant or are minor and due to normal chance causes. The chart can also show trends or wild and unpredictable fluctuations that could indicate readjustments in the process.

The $\overline{X}$- and R-charts are usually presented side by side because they reinforce each other in giving information about the distribution of the process. For a process to be operating under statistical control, points both in the $\overline{X}$- and R-charts should lie between the control limits. If all the points in both the charts remain within limits, the process capability is considered adequate. A lack of control

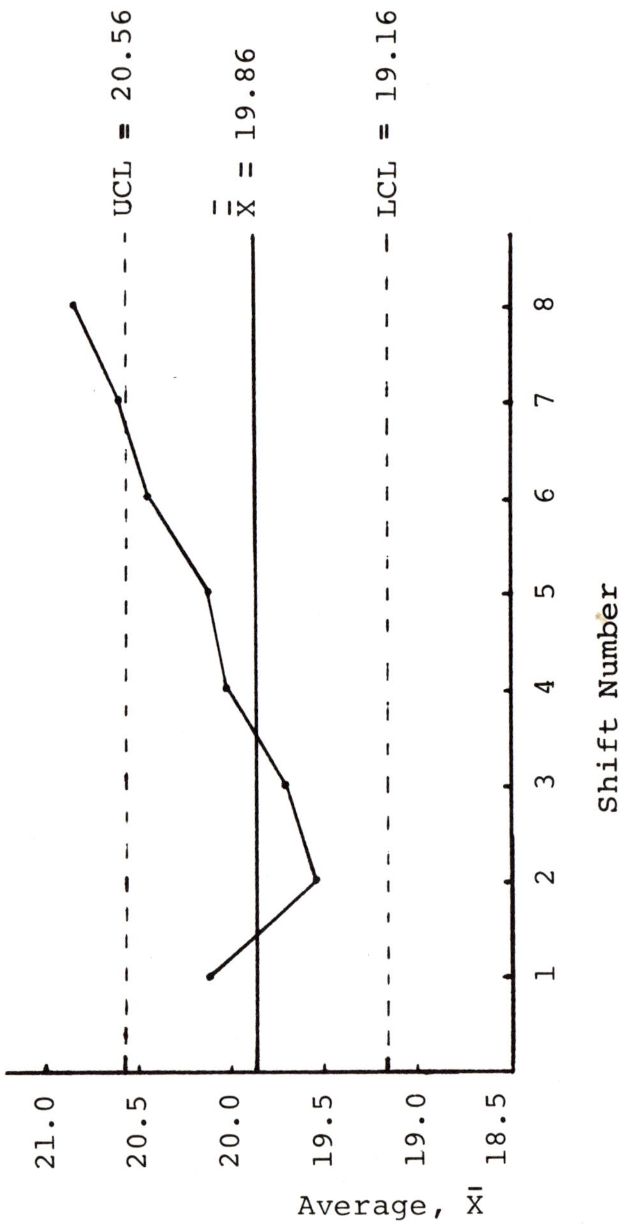

Fig. 13.7 $\overline{X}$*-chart for fill weight (fl. oz.).*

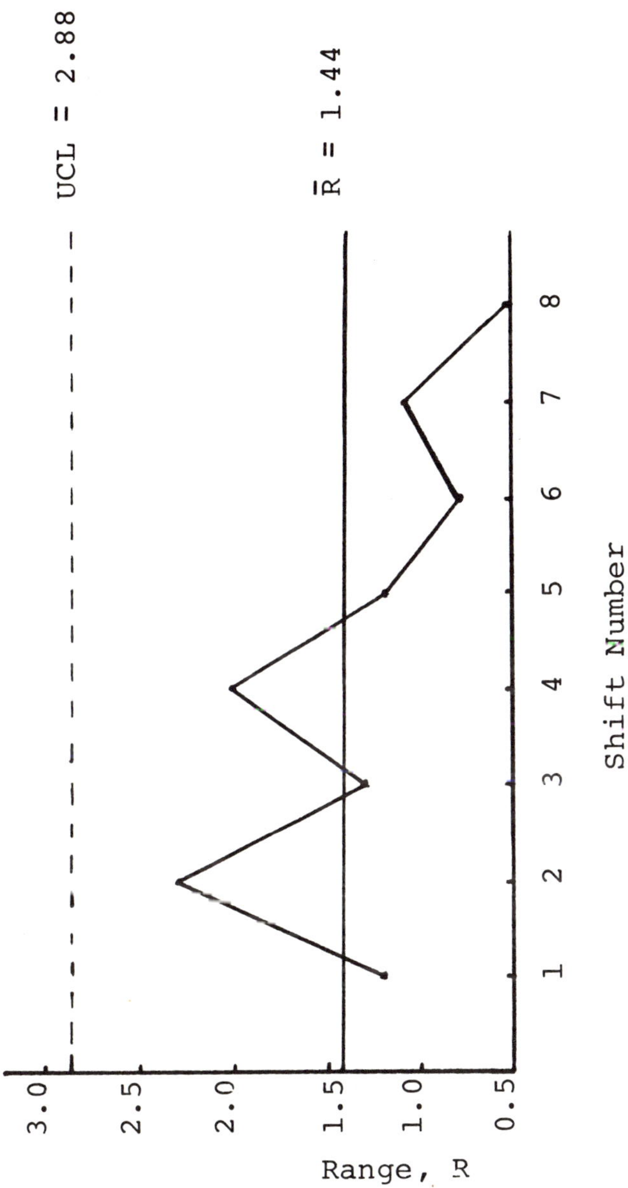

Fig. 13.8 *R-chart for fill weight (fl oz).*

is indicated in the following situations:

1. A point going outside the control limits is an indication of the presence of assignable causes.
2. A run of 7 or more points above or below the central line produces a pattern and indicates a shift in the process level. Sometimes a run of points beyond some secondary limits, for example, a run of 2 to 3 points beyond 2σ limits or a run of 4 to 5 points beyond 1σ limits or even a point in the vicinity of the control limits, is taken to be a warning of a possible lack of control or presence of nonrandom causes.

In such a case, some fundamental action has to be taken before control can be established. This action may be a review of specifications, or a change of method or material or even some of the characteristics of the process.

p-Charts

The p-chart is the most commonly used control chart method of analysis when the quality characteristic is an attribute, i.e., when the units are classified as defective or nondefective. Some examples of attribute measurements are percentage of dented cans, percent germination in seeds, or number of noxious weed seeds. The data generally follow a binomial distribution (discussed in chapter 5) with p′, the probability of success in a single trial, representing the probability of occurrence of a defective unit. The standard deviation of the fraction defective, p, in a sample of size n is

$\sigma_p = \sqrt{\dfrac{p'(1-p')}{n}}$, which is estimated by S_p. The p-chart is set as follows:

$$\text{center line} = \bar{p} = \text{mean fraction defective}$$

$$\text{UCL} = \bar{p} + 3\sqrt{\frac{\bar{p}(1-\bar{p})}{n}},$$

$$\text{and} \quad \text{LCL} = \bar{p} - 3\sqrt{\frac{\bar{p}(1-\bar{p})}{n}}$$

As with the $\overline{X}$-chart, it is important to establish control chart characteristics for the p-chart also. This may be accomplished by using past data, or a specification given by management, or by undertaking a special study.

The process is judged to be in statistical control in the same way as for $\overline{X}$- and R-charts. If all the sample points fall within the control limits without exhibiting any specific pattern, the process is said to be in control. In this case the average fraction defective $\overline{p}$ is taken as the standard fraction defective. If some points fall outside the control limits, a revised estimate of p' should be obtained by eliminating all the points that fall outside the limits. The *revised control limits* should be plotted and the process capability studied in relation to these revised limits.

As far as possible groups of equal sample size should be considered. This saves effort in making the control chart and makes the interpretation of the charts easier. If most of the samples are approximately equal in size, we can compute approximate control limits based on some typical sample size such as the average sample size or the most frequent sample size. Then, for any point near the limits, the exact limits based on the actual sample size for that point should be computed and plotted. Where the sample size varies from sample to sample, separate control limits are calculated for each sample. The smaller the sample size, the larger the control bands, and vice versa. If the sample size does not vary appreciably, a single set of control limits based on the average sample size can be used. Alternatively, for all sample sizes two sets of limits, one based on largest sample size and the other based on smallest sample size, can be constructed. Points falling outside the outer set of limits indicate the presence of assignable causes of variation which must be discovered and rectified; those within the inner set indicate that the process is in control. For all other points, action should be based on the exact control limits.

Example 13.4 A study was made to estimate the efficiency of a labeling machine putting labels on a canned food product. Samples of 50 cans were taken from each day's output and visually inspected to find out the number of cans wrongly labeled. The total number of wrongly labeled cans was counted for each day's output for 8 days and the data obtained are given in table 13.6.

Table 13.6 *Improperly labeled cans from example 13.4.*

Day	1	2	3	4	5	6	7	8	Total
No. of Cans Wrongly Labeled	3	4	12	14	2	4	3	1	43

$$\text{Mean fraction defective} = \bar{p} = \frac{43}{50 \times 8} = 0.11$$

Estimate of the standard deviation (σ_p)

$$= S_p = \sqrt{\frac{0.11(1 - 0.11)}{50}} = 0.044$$

Thus, for the p-chart,

$$\begin{aligned} \text{center line} &= \bar{p} = 0.11 \\ \text{UCL} &= \bar{p} + 3S_p = 0.11 + 3 \times 0.044 = 0.242 \\ \text{LCL} &= \bar{p} - 3S_p = 0.11 - 3 \times 0.044 \\ &= 0.0 \text{ (since LCL} < 0.0) \end{aligned}$$

Figure 13.9 shows the control chart and associated points for the 8 days of inspection.

At day 3, the machine is beginning to go out of control and by day 4 it is completely out of control. The cause of the problem was found and rectified and the machine showed a state of control afterwards.

To establish the final control limits that can be used for future studies of the process/machine capability, these p-chart limits are reset. The two points corresponding to days 3 and 4 are removed from the analysis and the control limits are recalculated as follows:

$$\Sigma_p = 17$$

$$\bar{p} = \frac{17}{6 \times 50} = 0.057$$

$$S_p = \sqrt{\frac{0.057(1 - 0.057)}{50}} = 0.033$$

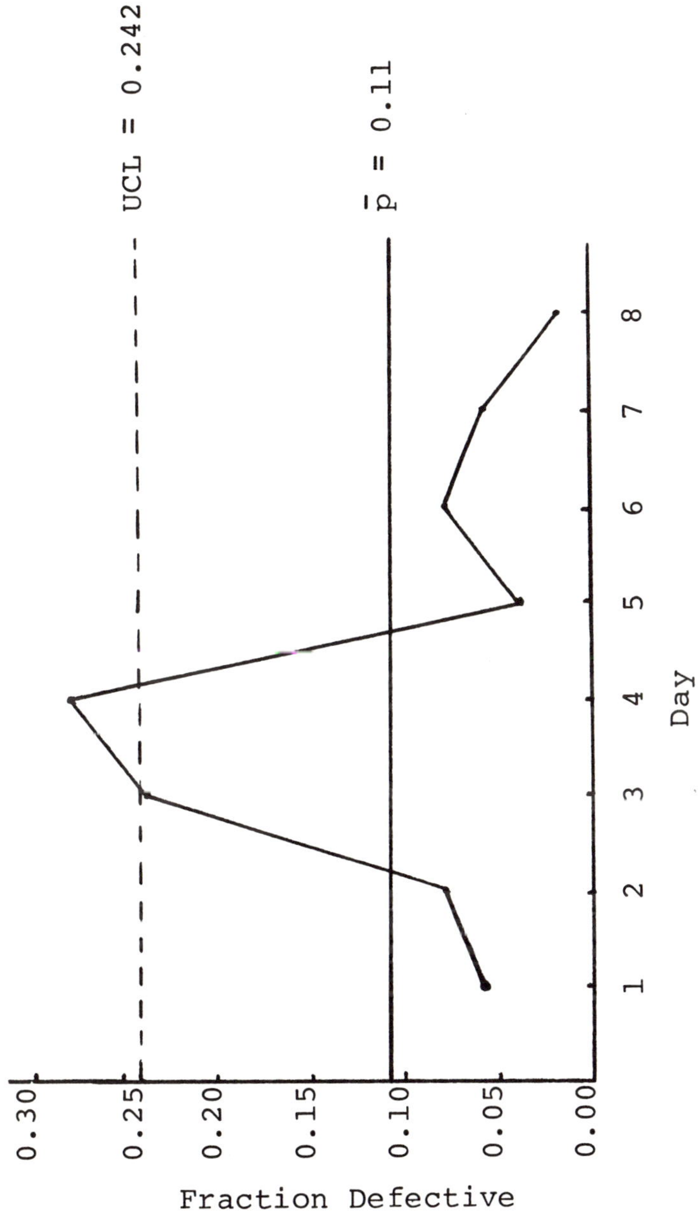

Fig. 13.9 *p-Chart for labeling operation.*

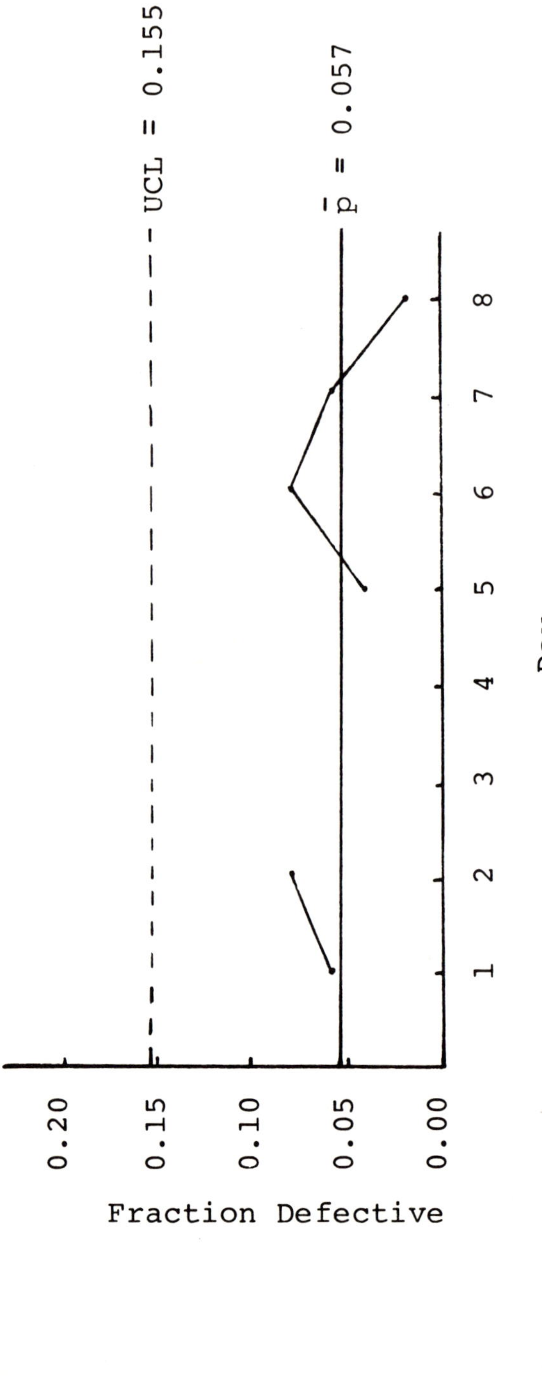

Fig. 13.10 *Revised p-chart for labeling operation.*

Thus, for the p-chart,

$$\begin{aligned}
\text{center line} &= \bar{p} = 0.057 \\
\text{UCL} &= \bar{p} + 3S_p = 0.057 + 3 \times 0.033 = 0.155 \\
\text{LCL} &= \bar{p} - 3S_p = 0.057 - 3 \times 0.033 \\
&= 0.0 \text{ (since LCL} < 0.0)
\end{aligned}$$

Figure 13.10 gives the new control limits with the data from 6 days (days 1, 2, 5, 6, 7, 8). As can be seen, these data are randomly distributed around the center line well within the control limits, exhibiting a state of statistical control for the machine.

Example 13.5 The following is a record of inspection of apples at a packing plant for a 1-year period, as reported by an inspector. The figure for each month is an average of several inspections over the month and the sample size is n = 100. The upper specification limit for apple quality is given as 7.5%.

Month	1	2	3	4	5	6	7	8	9	10	11	12
Average percentage defective	5.3	3.9	7.9	7.0	6.8	5.9	4.6	6.0	4.9	5.7	6.1	7.0

The calculations for the p-chart are as follows:

$$n = 100$$

$$\bar{p} = \frac{5.3 + 3.9 + \ldots + 7.0}{12} = 5.925\%$$

$$S_p = \sqrt{\frac{\bar{p}(1 - \bar{p})}{n}} = \sqrt{\frac{5.925(1 - 5.925)}{100}} = 2.361$$

$$\begin{aligned}
\text{center line} &= \bar{p} = 5.925 \\
\text{UCL} &= \bar{p} + 3S_p = 5.925 + 3 \times 2.361 = 13.008 \\
\text{LCL} &= \bar{p} - 3S_p = 5.925 - 3 \times 2.361 \\
&= 0.0 \text{ (since LCL} < 0.0)
\end{aligned}$$

Figure 13.11 shows the control limits and the inspection record for the 12 months.

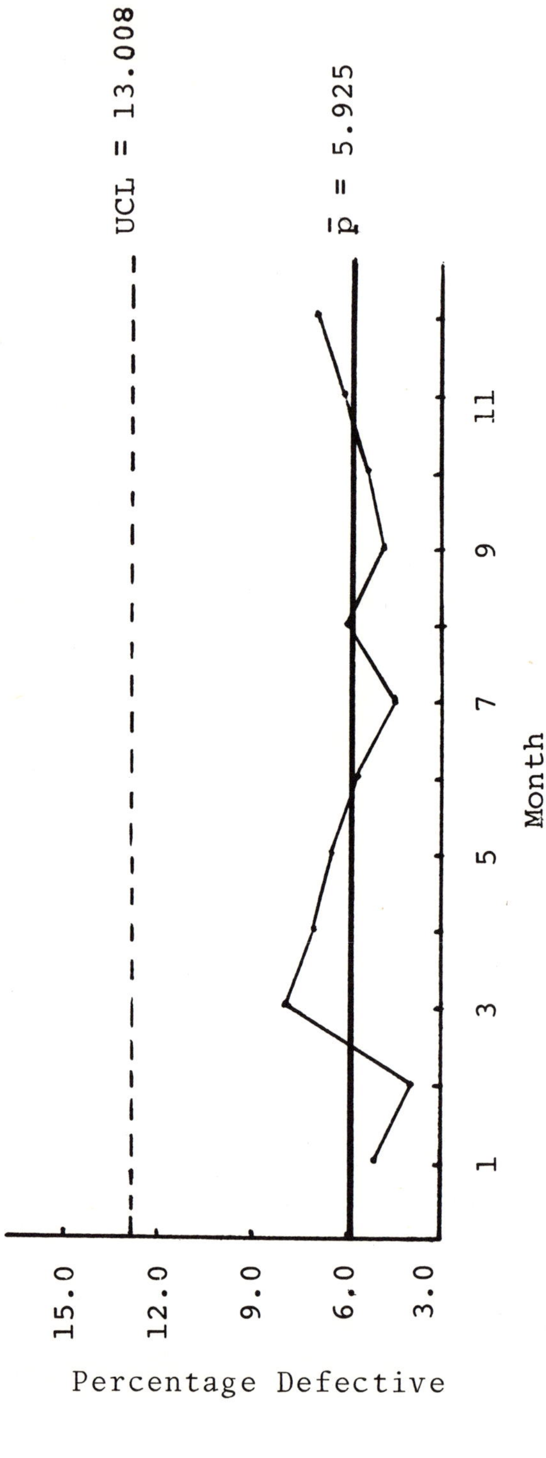

Figure 13.11 *p-Chart for percentage of defective apples.*

As can be seen from the figure, the apple quality is well within the control limits and usually also within the upper specification limit. Thus, a state of statistical control exists for the apple quality.

Exercises

1. Using Mil Std 105D, derive a single sampling plan for a lot size of 17000, an AQL of 2.5% under normal inspection conditions.

2. Derive a single sampling plan with the following specifications:
   acceptable quality level (AQL) = 2.5%
   producer's risk (α) = 0.05
   lot tolerance percent defective (LTPD) = 6.0%
   consumer's acceptance probability (β) = 0.10

3. Plot an OC curve for the sampling plan of exercise 2 and find the probability of acceptance of a lot that is 3% defective.

4. Determine the probabilities of acceptance and plot the OC curve when n = 195 and c = 3.

5. Using Mil Std 105D, derive a single sampling plan for a lot size of 9000, an AQL of 1.0%, and inspection level II (normal inspection). Also, determine the probabilities of acceptance and plot the OC curve for the sampling plan. What would be the probability of acceptance of a lot that is 3.0% defective under this sampling plan and OC curve?

6. A Department of Agriculture inspector has been using a single sampling plan with n = 50, acceptance number (c) = 5, and rejection number (r) = 6 for inspecting heads of lettuce. This sampling plan was derived from Mil Std 105D for a lot size of 400 and an AQL of 4.0%. This plan has been used for a long time and it is now believed that the product capability is running at 5.0% defective. Calculate the producer's risk for percent defective at 5.0% using the OC curve drawn with AQL of 4.0%, n = 50, and c = 5.

7. The following data record the gross weight (in oz) of cans of pears produced in a cannery. A sample of 5 cans was taken each shift for a total of 10 shift periods. The gross weight of each can of pears is supposed to be 25.5 oz.

Shift	Gross weight of cans of pears (oz)				
1	26.5	27.2	26.9	28.2	25.7
2	24.9	26.2	28.3	27.0	27.3
3	27.4	26.8	28.0	25.9	25.0
4	25.6	25.9	26.1	27.2	26.8
5	26.7	27.1	28.2	25.3	24.9
6	27.2	25.6	25.0	25.7	26.2
7	25.9	25.1	25.7	26.2	27.0
8	29.7	29.6	30.1	32.4	31.7
9	32.0	33.1	32.9	33.0	32.9
10	33.0	32.9	32.7	33.4	32.6

Process capability studies, made in the past, show the process average at 26.2 oz with mean range at 0.47 oz. Construct an $\overline{X}$-chart and an R-chart and plot the data from the ten shifts. Discuss the results.

8. A quality control supervisor wishes to set up control charts for a continuous process. The supervisor takes samples of size 8 and computes the mean and range for each sample. The grand mean of all the sample means is $\overline{X} = 16.7$ oz and the mean of the ranges is $\overline{R} = 0.94$. Set up the $\overline{X}$-chart and the R-chart for the process.

9. Samples of 100 items are taken from each of 15 lots and inspected. This is the number of items rejected from each batch.

Lot	1	2	3	4	5	6	7	8	9	10	11	12	13	14	15
No. of items rejected	5	7	11	10	9	12	13	10	11	9	8	7	10	12	13

Set up a p-chart for the fraction defective.

10. Derive a single sampling plan for which AQL = 0.02, α = 0.05, LTPD = 0.14, and β = 0.10.

11. Using Mil Std 105D, determine the sample size and acceptance number for a lot size of 4000, an AQL of 2.5% for an inspection level of S-III (a destructive sampling situation).

12. Plot an OC curve for n = 450 and c = 2. What is the probability of acceptance of a lot that is 4% defective?

13. The inspection records of 10 inspections, each of sample size 75, of a poultry product in a period of one month show the following percent defective:

Sample	1	2	3	4	5	6	7	8	9	10
Percentage defective	3.4	3.5	4.1	4.3	3.9	3.1	4.4	4.6	4.0	3.7

Set up a p-chart for the data.

14. The following data give the mean weight (in gm) and ranges of a sample of 100 apples picked randomly from 10 lots.

Lot	1	2	3	4	5	6	7	8	9	10
Mean	15.6	12.8	9.3	10.6	8.2	11.5	10.0	9.5	12.6	11.2
Range	0.72	0.79	0.81	0.69	0.91	0.66	0.75	0.89	0.92	0.90

Set up an $\overline{X}$-chart and an R-chart for the data.

Selected Bibliography

A. Some General Texts on Applied Statistics

1. Brownlee, K. A. *Statistical theory and methodology in science and engineering.* New York: John Wiley and Sons, Inc., 1965.
2. Cochran, W. G., and Cox, G. M. *Experimental design.* New York: John Wiley and Sons, Inc., 1957.
3. Cox, D. R. *Planning of experiments.* New York: John Wiley and Sons, Inc., 1958.
4. Draper, N. R., and Smith, H. *Applied regression analysis.* New York: John Wiley and Sons, Inc., 1968.
5. Malik, H. J. and Mullen, K. *A first course in probability and statistics.* Reading, Massachusetts: Addison-Wesley Publishing Co., 1971.
6. Malik, H. J., and Mullen, K. *Applied statistics for business and economics.* Reading, Massachusetts: Addison-Wesley Publishing Co., 1975.
7. Maxwell, A. E. *Analysing qualitative data.* London: Methuen and Co. Ltd., 1961.
8. Mendenhall, W. *Introduction to probability and statistics.* Belmont, California: Wadsworth Publishing Co., Inc., 1971.
9. Remington, R. D., and Schork, M. A. *Statistics with applications to the biological and health sciences.* Englewood Cliffs, New Jersey: Prentice-Hall, Inc., 1970.
10. Rickmars, A. D., and Todd, H. N. *Statistics: an introduction.* New York: McGraw-Hill Book Co., 1967.
11. Sokal, R. R., and Rohlf, F. J. *Biometry: the principles and practice of statistics in biological research.* San Francisco: W. H. Freeman and Company, 1969.

12. Steel, R. G. D., and Torrie, J. H. *Principles and procedures of statistics.* New York: McGraw-Hill Book Co., 1960.

B. Some Texts on Nonparametric Statistics

1. Conover, W. J. *Practical nonparametric statistics.* New York: John Wiley and Sons, Inc., 1971.
2. Pierce, A. *Fundamentals of nonparametric statistics.* Belmont, California: Dickenson Publishing Company, Inc., 1970.
3. Siegel, S. *Nonparametric statistics.* New York: McGraw-Hill Book Co., 1956.

C. Some Texts on Quality Control

1. Duncan, A. J. *Quality control and industrial statistics.* Homewood, Illinois: Richard D. Irwin, Inc. 1974.
2. Grant, E. L., and Leavenworth, R. S., *Statistical quality control.* New York: McGraw-Hill Book Co., 1972.
3. Hansen, B. L. *Quality control: theory and applications.* Englewood Cliffs, New Jersey: Prentice-Hall, Inc., 1963.
4. Juran, J. M., and Gryna, F. M., Jr. *Quality planning and analysis.* New York: McGraw-Hill Book Co., 1970.
5. Kramer, A., and Twigg, B. A. *Quality control for the food industry.* Westport, Connecticut: The AVI Publishing Company, 1970.
6. Puri, S. C., Ennis, D., and Mullen, K. *Statistical quality control for food and agricultural scientists.* Boston, Massachusetts: G. K. Hall and Co., 1979.
7. Shewhart, W. A. *Statistical methods from the viewpoint of quality control.* Washington, D.C.: Department of Agriculture, 1939.

D. Some Texts on Sampling Theory

1. Cochran, W. G. *Sampling techniques.* New York: John Wiley and Sons, Inc., 1963.
2. Deming, W. E. *Sample design in business research.* New York: John Wiley and Sons, Inc., 1960.
3. Kish, L. *Survey sampling.* New York: John Wiley and Sons, Inc., 1965.
4. Mendenhall, W., Ott, L., and Scheaffer, R. L. *Elementary survey sampling.* Belmont, California: Wadsworth Publishing Company, Inc., 1971.

E. Some Statistical Tables

1. Dodge, H. F., and Romig, H. G. *Sampling inspection tables. Single and double sampling.* New York: John Wiley and Sons, Inc., 1959.
2. Fisher, R. A., and Yates, F. *Statistical tables for biological, agricultural, and medical research.* (6th ed.) Harlow, Essex, England: Longman Group Limited, 1975.
3. Pearson, E. S., and Hartley, H. O. *Biometrika tables for statisticians.* vol. 1. New York: Cambridge University Press, 1966.
4. Rand Corporation. *A million random digits with 100,000 normal deviates.* Glencoe, Illinois: The Free Press, 1955.
5. United States Department of Defense, Military Standards: MIL. STD. 105D. *Sampling procedures and tables for inspection by attributes.* Washington, D.C.: U.S. Government Printing Office, 1963.

A

Appendix

Table 1. *Binomial Coefficients* $\binom{n}{x} = \frac{n!}{x!(n - x)!}$.

	x										
n	0	1	2	3	4	5	6	7	8	9	10
0	1										
1	1	1									
2	1	2	1								
3	1	3	3	1							
4	1	4	6	4	1						
5	1	5	10	10	5	1					
6	1	6	15	20	15	6	1				
7	1	7	21	35	35	21	7	1			
8	1	8	28	56	70	56	28	8	1		
9	1	9	36	84	126	126	84	36	9	1	
10	1	10	45	120	210	252	210	120	45	10	1
11	1	11	55	165	330	462	462	330	165	55	11
12	1	12	66	220	495	792	924	792	495	220	66
13	1	13	78	286	715	1287	1716	1716	1287	715	286
14	1	14	91	364	1001	2002	3003	3432	3003	2002	1001
15	1	15	105	455	1365	3003	5005	6435	6435	5005	3003
16	1	16	120	560	1820	4368	8008	11440	12870	11440	8008
17	1	17	136	680	2380	6188	12376	19448	24310	24310	19448
18	1	18	153	816	3060	8568	18564	31824	43758	48620	43758
19	1	19	171	969	3876	11628	27132	50388	75582	92378	92378
20	1	20	190	1140	4845	15504	38760	77520	125970	167960	184756

Table 2. *Values of Negative Exponential Function* e^{-x}.

x	e^{-x}	*x*	e^{-x}	*x*	e^{-x}	*x*	e^{-x}	*x*	e^{-x}
0.0	1.000	2.0	0.135	4.0	0.018	6.0	0.0025	8.0	0.00034
0.1	0.905	2.1	0.122	4.1	0.017	6.1	0.0022	8.1	0.00030
0.2	0.819	2.2	0.111	4.2	0.015	6.2	0.0020	8.2	0.00028
0.3	0.741	2.3	0.100	4.3	0.014	6.3	0.0018	8.3	0.00025
0.4	0.670	2.4	0.091	4.4	0.012	6.4	0.0017	8.4	0.00023
0.5	0.607	2.5	0.082	4.5	0.011	6.5	0.0015	8.5	0.00020
0.6	0.549	2.6	0.074	4.6	0.010	6.6	0.0014	8.6	0.00018
0.7	0.497	2.7	0.067	4.7	0.009	6.7	0.0012	8.7	0.00017
0.8	0.449	2.8	0.061	4.8	0.008	6.8	0.0011	8.8	0.00015
0.9	0.407	2.9	0.055	4.9	0.007	6.9	0.0010	8.9	0.00014
1.0	0.368	3.0	0.050	5.0	0.0067	7.0	0.0009	9.0	0.00012
1.1	0.333	3.1	0.045	5.1	0.0061	7.1	0.0008	9.1	0.00011
1.2	0.301	3.2	0.041	5.2	0.0055	7.2	0.0007	9.2	0.00010
1.3	0.273	3.3	0.037	5.3	0.0050	7.3	0.0007	9.3	0.00009
1.4	0.247	3.4	0.033	5.4	0.0045	7.4	0.0006	9.4	0.00008
1.5	0.223	3.5	0.030	5.5	0.0041	7.5	0.00055	9.5	0.00008
1.6	0.202	3.6	0.027	5.6	0.0037	7.6	0.00050	9.6	0.00007
1.7	0.183	3.7	0.025	5.7	0.0033	7.7	0.00045	9.7	0.00006
1.8	0.165	3.8	0.022	5.8	0.0030	7.8	0.00041	9.8	0.00006
1.9	0.150	3.9	0.020	5.9	0.0027	7.9	0.00037	9.9	0.00005

Table 3. *Normal Curve Areas.*

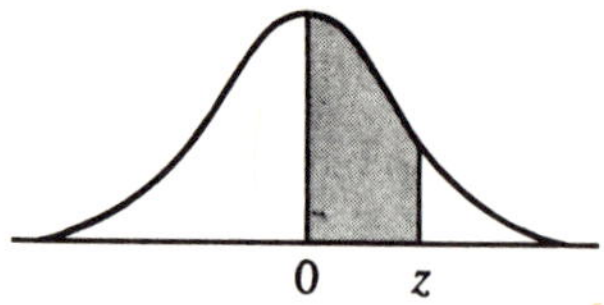

z	.00	.01	.02	.03	.04	.05	.06	.07	.08	.09
0.0	.0000	.0040	.0080	.0120	.0160	.0199	.0239	.0279	.0319	.0359
0.1	.0398	.0438	.0478	.0517	.0557	.0596	.0636	.0675	.0714	.0753
0.2	.0793	.0832	.0871	.0910	.0948	.0987	.1026	.1064	.1103	.1141
0.3	.1179	.1217	.1255	.1293	.1331	.1368	.1406	.1443	.1480	.1517
0.4	.1554	.1591	.1628	.1664	.1700	.1736	.1772	.1808	.1844	.1879
0.5	.1915	.1950	.1985	.2019	.2054	.2088	.2123	.2157	.2190	.2224
0.6	.2257	.2291	.2324	.2357	.2389	.2422	.2454	.2486	.2517	.2549
0.7	.2580	.2611	.2642	.2673	.2704	.2734	.2764	.2794	.2823	.2852
0.8	.2881	.2910	.2939	.2967	.2995	.3023	.3051	.3078	.3106	.3133
0.9	.3159	.3186	.3212	.3238	.3264	.3289	.3315	.3340	.3365	.3389
1.0	.3413	.3438	.3461	.3485	.3508	.3531	.3554	.3577	.3599	.3621
1.1	.3643	.3665	.3686	.3708	.3729	.3749	.3770	.3790	.3810	.3830
1.2	.3849	.3869	.3888	.3907	.3925	.3944	.3962	.3980	.3997	.4015
1.3	.4032	.4049	.4066	.4082	.4099	.4115	.4131	.4147	.4162	.4177
1.4	.4192	.4207	.4222	.4236	.4251	.4265	.4279	.4292	.4306	.4319
1.5	.4332	.4345	.4357	.4370	.4382	.4394	.4406	.4418	.4429	.4441
1.6	.4452	.4463	.4474	.4484	.4495	.4505	.4515	.4525	.4535	.4545
1.7	.4554	.4564	.4573	.4582	.4591	.4599	.4608	.4616	.4625	.4633
1.8	.4641	.4649	.4656	.4664	.4671	.4678	.4686	.4693	.4699	.4706
1.9	.4713	.4719	.4726	.4732	.4738	.4744	.4750	.4756	.4761	.4767
2.0	.4772	.4778	.4783	.4788	.4793	.4798	.4803	.4808	.4812	.4817
2.1	.4821	.4826	.4830	.4834	.4838	.4842	.4846	.4850	.4854	.4857
2.2	.4861	.4864	.4868	.4871	.4875	.4878	.4881	.4884	.4887	.4890
2.3	.4893	.4896	.4898	.4901	.4904	.4906	.4909	.4911	.4913	.4916
2.4	.4918	.4920	.4922	.4925	.4927	.4929	.4931	.4932	.4934	.4936
2.5	.4938	.4940	.4941	.4943	.4945	.4946	.4948	.4949	.4951	.4952
2.6	.4953	.4955	.4956	.4957	.4959	.4960	.4961	.4962	.4963	.4964
2.7	.4965	.4966	.4967	.4968	.4969	.4970	.4971	.4972	.4973	.4974
2.8	.4974	.4975	.4976	.4977	.4977	.4978	.4979	.4979	.4980	.4981
2.9	.4981	.4982	.4982	.4983	.4984	.4984	.4985	.4985	.4986	.4986
3.0	.4987	.4987	.4987	.4988	.4988	.4989	.4989	.4989	.4990	.4990

Table 4. *Percentage Points of the t-Distribution.*

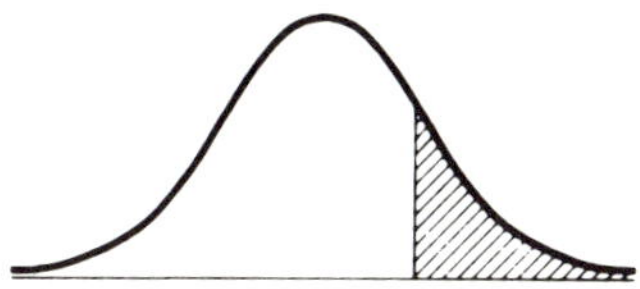

t_a

d. f.	$t_{.40}$	$t_{.30}$	$t_{.20}$	$t_{.10}$	$t_{.05}$	$t_{.025}$	$t_{.01}$	$t_{.005}$	$t_{.0005}$
1	0.3250	0.7270	1.376	3.078	6.3138	12.706	31.821	63.657	636.619
2	.2885	.6172	1.061	1.886	2.9200	4.3027	6.965	9.9248	31.598
3	.2766	.5840	.978	1.638	2.3534	3.1825	4.541	5.8409	12.924
4	.2707	.5692	.941	1.533	2.1318	2.7764	3.747	4.6041	8.610
5	.2672	.5598	.920	1.476	2.0150	2.5706	3.365	4.0321	6.869
6	.2648	.5536	.906	1.440	1.9432	2.4469	3.143	3.7074	5.959
7	.2632	.5493	.896	1.415	1.8946	2.3646	2.998	3.4995	5.408
8	.2619	.5461	.889	1.397	1.8595	2.3060	2.896	3.3554	5.041
9	.2610	.5436	.883	1.383	1.8331	2.2622	2.821	3.2498	4.781
10	.2602	.5416	.879	1.372	1.8125	2.2281	2.764	3.1693	4.587
11	.2596	.5400	.876	1.363	1.7939	2.2010	2.718	3.1058	4.437
12	.2590	.5387	.873	1.356	1.7823	2.1788	2.681	3.0545	4.318
13	.2586	.5375	.870	1.350	1.7709	2.1604	2.650	3.0123	4.221
14	.2582	.5366	.868	1.345	1.7613	2.1448	2.624	2.9768	4.140
15	.2579	.5358	.866	1.341	1.7530	2.1315	2.602	2.9467	4.073
16	.2576	.5351	.865	1.337	1.7459	2.1199	2.583	2.9208	4.015
17	.2574	.5344	.863	1.333	1.7396	2.1098	2.567	2.8982	3.965
18	.2571	.5338	.862	1.330	1.7341	2.1009	2.552	2.8784	3.922
19	.2569	.5333	.861	1.328	1.7291	2.0930	2.539	2.8609	3.883
20	.2567	.5329	.860	1.325	1.7247	2.0860	2.528	2.8453	3.850
21	.2566	.5325	.859	1.323	1.7207	2.0796	2.518	2.8314	3.810
22	.2564	.5321	.858	1.321	1.7171	2.0739	2.508	2.8188	3.792
23	.2563	.5318	.858	1.319	1.7139	2.0687	2.500	2.9073	3.767
24	.2562	.5315	.857	1.318	1.7109	2.0639	2.492	2.7969	3.745
25	.2561	.5312	.856	1.316	1.7081	2.0595	2.485	2.7874	3.725
26	.2560	.5309	.856	1.315	1.7056	2.0555	2.479	2.7787	3.707
27	.2559	.5307	.855	1.314	1.7033	2.0518	2.473	2.7707	3.690
28	.2558	.5304	.855	1.313	1.7011	2.0484	2.467	2.7633	3.674
29	.2557	.5302	.854	1.311	1.6991	2.0452	2.462	2.7564	3.659
30	.2556	.5300	.854	1.310	1.6973	2.0423	2.457	2.7500	3.616
35	.2553	.5292	.8521	1.3062	1.6896	2.0301	2.438	2.7239	3.5919
40	.2550	.5286	.8507	1.3031	1.6839	2.0211	2.423	2.7045	3.5511
45	.2549	.5281	.8497	1.3007	1.6794	2.0141	2.412	2.6896	3.5207
50	.2547	.5278	.8489	1.2987	1.6759	2.0086	2.403	2.6778	3.4965
60	.2545	.5272	.8477	1.2959	1.6707	2.0003	2.390	2.6603	3.4606
70	.2543	.5268	.8468	1.2938	1.6669	1.9945	2.381	2.6480	3.4355
80	.2542	.5265	.8462	1.2922	1.6641	1.9901	2.374	2.6388	3.4169
90	.2541	.5263	.8457	1.2910	1.6620	1.9867	2.368	2.6316	3.4022
100	.2540	.5261	.8452	1.2901	1.6602	1.9840	2.364	2.6260	3.3909
120	.2539	.5258	.8446	1.2887	1.6577	1.9799	2.358	2.6175	3.3736
140	.2538	.5256	.8442	1.2876	1.6558	1.9771	2.353	2.6114	3.3615
160	.2538	.5255	.8439	1.2869	1.6545	1.9749	2.350	2.6070	3.3527
180	.2537	.5253	.8436	1.2863	1.6534	1.9733	2.347	2.6035	3.3456
200	.2537	.5252	.8434	1.2858	1.6525	1.9719	2.345	2.6006	3.3400
∞	.2533	.5244	.8416	1.2816	1.6449	1.9600	2.326	2.5758	3.2905

SOURCE: Reproduced from *Documenta Geigy Scientific Tables,* 7th edition, by permission of CIBA-GEIGY Limited, Basle, Switzerland.

Table 5. *Percentage Points of the* χ^2*-Distribution.*

ν \ Q	0·995	0·990	0·975	0·950	0·900	0·750	0·500
1	$392704 . 10^{-10}$	$157088 . 10^{-9}$	$982069 . 10^{-9}$	$393214 . 10^{-8}$	0·0157908	0·1015308	0·454937
2	0·0100251	0·0201007	0·0506356	0·102587	0·210720	0·575364	1·38629
3	0·0717212	0·114832	0·215795	0·351846	0·584375	1·212534	2·36597
4	0·206990	0·297110	0·484419	0·710721	1·063623	1·92255	3·35670
5	0·411740	0·554300	0·831211	1·145476	1·61031	2·67460	4·35146
6	0·675727	0·872085	1·237347	1·63539	2·20413	3·45460	5·34812
7	0·989265	1·239043	1·68987	2·16735	2·83311	4·25485	6·34581
8	1·344419	1·646482	2·17973	2·73264	3·48954	5·07064	7·34412
9	1·734926	2·087912	2·70039	3·32511	4·16816	5·89883	8·34283
10	2·15585	2·55821	3·24697	3·94030	4·86518	6·73720	9·34182
11	2·60321	3·05347	3·81575	4·57481	5·57779	7·58412	10·3410
12	3·07382	3·57056	4·40379	5·22603	6·30380	8·43842	11·3403
13	3·56503	4·10691	5·00874	5·89186	7·04150	9·29906	12·3398
14	4·07468	4·66043	5·62872	6·57063	7·78953	10·1653	13·3393
15	4·60094	5·22935	6·26214	7·26094	8·54675	11·0365	14·3389
16	5·14224	5·81221	6·90766	7·96164	9·31223	11·9122	15·3385
17	5·69724	6·40776	7·56418	8·67176	10·0852	12·7919	16·3381
18	6·26481	7·01491	8·23075	9·39046	10·8649	13·6753	17·3379
19	6·84398	7·63273	8·90655	10·1170	11·6509	14·5620	18·3376

20	7·43386	8·26040	9·59083	10·8508	12·4426	15·4518	19·3374
21	8·03366	8·89720	10·28293	11·5913	13·2396	16·3444	20·3372
22	8·64272	9·54249	10·9823	12·3380	14·0415	17·2396	21·3370
23	9·26042	10·19567	11·6885	13·0905	14·8479	18·1373	22·3369
24	9·88623	10·8564	12·4011	13·8484	15·6587	19·0372	23·3367
25	10·5197	11·5240	13·1197	14·6114	16·4734	19·9393	24·3366
26	11·1603	12·1981	13·8439	15·3791	17·2919	20·8434	25·3364
27	11·8076	12·8786	14·5733	16·1513	18·1138	21·7494	26·3363
28	12·4613	13·5648	15·3079	16·9279	18·9392	22·6572	27·3363
29	13·1211	14·2565	16·0471	17·7083	19·7677	23·5666	28·3362
30	13·7867	14·9535	16·7908	18·4926	20·5992	24·4776	29·3360
40	20·7065	22·1643	24·4331	26·5093	29·0505	33·6603	39·3354
50	27·9907	29·7067	32·3574	34·7642	37·6886	42·9421	49·3349
60	35·5346	37·4848	40·4817	43·1879	46·4589	52·2938	59·3347
70	43·2752	45·4418	48·7576	51·7393	55·3290	61·6983	69·3344
80	51·1720	53·5400	57·1532	60·3915	64·2778	71·1445	79·3343
90	59·1963	61·7541	65·6466	69·1260	73·2912	80·6247	89·3342
100	67·3276	70·0648	74·2219	77·9295	82·3581	90·1332	99·3341
X	−2·5758	−2·3263	−1·9600	−1·6449	−1·2816	−0·6745	0·0000

Source: Pearson and Hartley, 1966.
Note: Abridged with permission from *Biometrika Tables for Statisticians* Vol. 1.

Table 5. *(continued)*

ν \ Q	0·250	0·100	0·050	0·025	0·010	0·005	0·001
1	1·32330	2·70554	3·84146	5·02389	6·63490	7·87944	10·828
2	2·77259	4·60517	5·99147	7·37776	9·21034	10·5966	13·816
3	4·10835	6·25139	7·81473	9·34840	11·3449	12·8381	16·266
4	5·38527	7·77944	9·48773	11·1433	13·2767	14·8602	18·467
5	6·62568	9·23635	11·0705	12·8325	15·0863	16·7496	20·515
6	7·84080	10·6446	12·5916	14·4494	16·8119	18·5476	22·458
7	9·03715	12·0170	14·0671	16·0128	18·4753	20·2777	24·322
8	10·2188	13·3616	15·5073	17·5346	20·0902	21·9550	26·125
9	11·3887	14·6837	16·9190	19·0228	21·6660	23·5893	27·877
10	12·5489	15·9871	18·3070	20·4831	23·2093	25·1882	29·588
11	13·7007	17·2750	19·6751	21·9200	24·7250	26·7569	31·264
12	14·8454	18·5494	21·0261	23·3367	26·2170	28·2995	32·909
13	15·9839	19·8119	22·3621	24·7356	27·6883	29·8194	34·528
14	17·1170	21·0642	23·6848	26·1190	29·1413	31·3193	36·123
15	18·2451	22·3072	24·9958	27·4884	30·5779	32·8013	37·697
16	19·3688	23·5418	26·2962	28·8454	31·9999	34·2672	39·252
17	20·4887	24·7690	27·5871	30·1910	33·4087	35·7185	40·790
18	21·6049	25·9894	28·8693	31·5264	34·8053	37·1564	42·312
19	22·7178	27·2036	30·1435	32·8523	36·1908	38·5822	43·820

20	23·8277	28·4120	31·4104	34·1696	37·5662	39·9968	45·315
21	24·9348	29·6151	32·6705	35·4789	38·9321	41·4010	46·797
22	26·0393	30·8133	33·9244	36·7807	40·2894	42·7956	48·268
23	27·1413	32·0069	35·1725	38·0757	41·6384	44·1813	49·728
24	28·2412	33·1963	36·4151	39·3641	42·9798	45·5585	51·179
25	29·3389	34·3816	37·6525	40·6465	44·3141	46·9278	52·620
26	30·4345	35·5631	38·8852	41·9232	45·6417	48·2899	54·052
27	31·5284	36·7412	40·1133	43·1944	46·9630	49·6449	55·476
28	32·6205	37·9159	41·3372	44·4607	48·2782	50·9933	56·892
29	33·7109	39·0875	42·5569	45·7222	49·5879	52·3356	58·302
30	34·7998	40·2560	43·7729	46·9792	50·8922	53·6720	59·703
40	45·6160	51·8050	55·7585	59·3417	63·6907	66·7659	73·402
50	56·3336	63·1671	67·5048	71·4202	76·1539	79·4900	86·661
60	66·9814	74·3970	79·0819	83·2976	88·3794	91·9517	99·607
70	77·5766	85·5271	90·5312	95·0231	100·425	104·215	112·317
80	88·1303	96·5782	101·879	106·629	112·329	116·321	124·839
90	98·6499	107·565	113·145	118·136	124·116	128·299	137·208
100	109·141	118·498	124·342	129·561	135·807	140·169	149·449
X	+0·6745	+1·2816	+1·6449	+1·9600	+2·3263	+2·5758	+3·0902

Table 6.1. *Binomial Confidence Intervals with 1 − α = 0.95.*

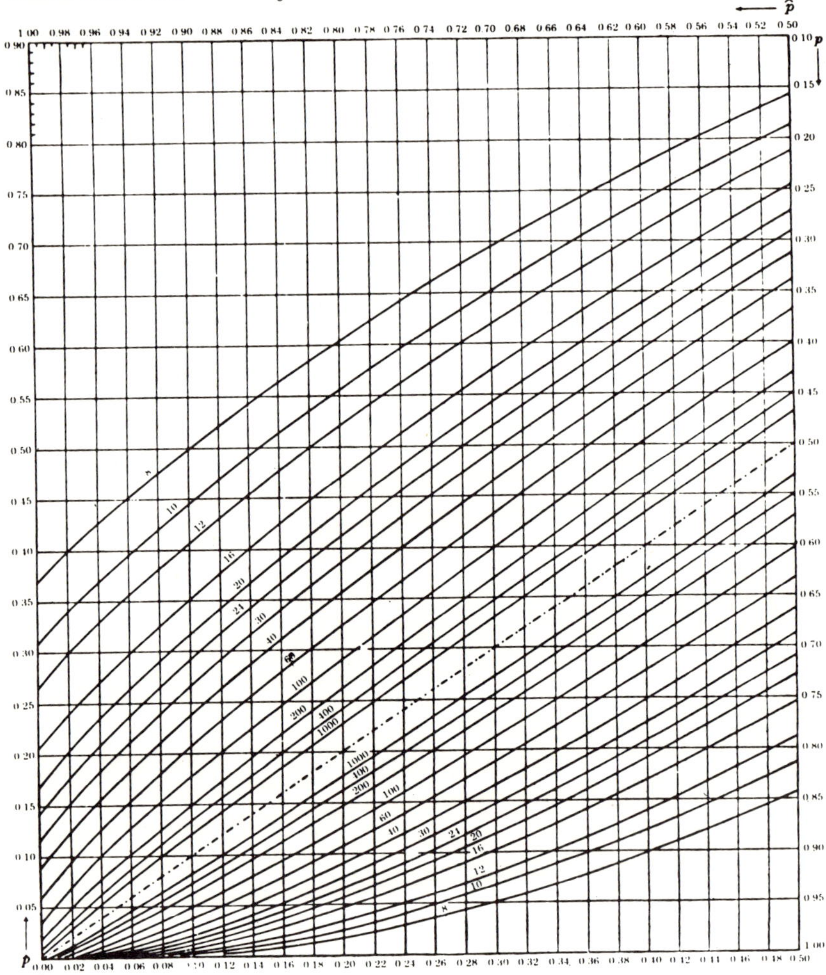

SOURCE: Pearson and Hartley, 1966.
NOTE: The numbers printed along the curves indicate the sample size n. If for a given value of the abscissa $\hat{p}$, p_A and p_B are the ordinates read from (or interpolated between) the appropriate lower and upper curves, then

$$Pr\{p_A < p < p_B\} < 1 - 2\alpha.$$

Table 6.2. *Binomial Confidence Intervals with 1 − α = 0.99.*

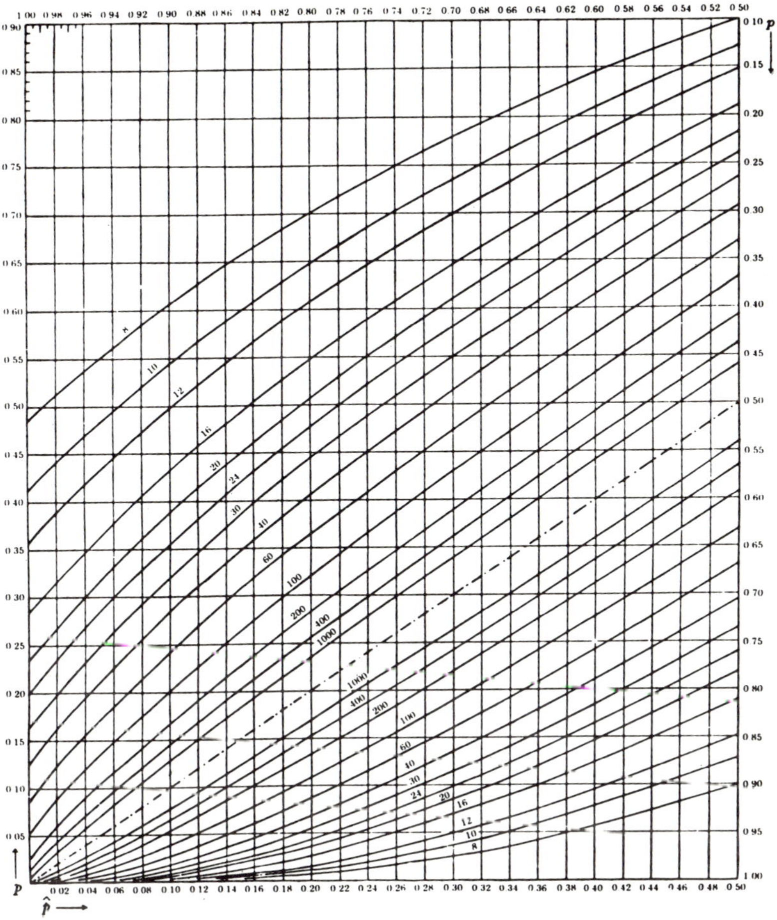

SOURCE: Pearson and Hartley, 1966.
NOTE: The process of reading from the curves can be simplified with the help of the right-angled corner of a loose sheet of paper or thin card, along the edges of which are marked off the scales shown in the top left-hand corner of each chart.
The numbers printed along the curves indicate the sample size n.

Table 7. *Percentiles of the Skewness Test Statistic* a_3.

Size of sample n	$(a_3)_{0.95}$	$(a_3)_{0.99}$	Size of sample n	$(a_3)_{0.95}$	$(a_3)_{0.99}$
25	0.711	1.061	200	0.280	0.403
30	0.662	0.986	250	0.251	0.360
35	0.621	0.923	300	0.230	0.329
40	0.587	0.870	350	0.213	0.305
45	0.558	0.825	400	0.200	0.285
50	0.534	0.787	450	0.188	0.269
			500	0.179	0.255
60	0.492	0.723	550	0.171	0.243
70	0.459	0.673	600	0.163	0.233
80	0.432	0.631	650	0.157	0.224
90	0.409	0.596	700	0.151	0.215
100	0.389	0.567	750	0.146	0.208
			800	0.142	0.202
125	0.350	0.508	850	0.138	0.196
150	0.321	0.464	900	0.134	0.190
175	0.298	0.430	950	0.130	0.185
200	0.280	0.403	1000	0.127	0.180

SOURCE: Pearson and Hartley, 1966.

Table 8. *Percentiles of Geary's Kurtosis Test Statistic g.*

Size of sample n	$g_{0.01}$	$g_{0.05}$	$g_{0.10}$	$g_{0.90}$	$g_{0.95}$	$g_{0.99}$
11	0.6675	0.7153	0.7400	0.8899	0.9073	0.9359
16	0.6829	0.7236	0.7452	0.8733	0.8884	0.9137
21	0.6950	0.7304	0.7495	0.8631	0.8768	0.9001
26	0.7040	0.7360	0.7530	0.8570	0.8686	0.8901
31	0.7110	0.7404	0.7559	0.8511	0.8625	0.8827
36	0.7167	0.7440	0.7583	0.8468	0.8578	0.8769
41	0.7216	0.7470	0.7604	0.8436	0.8540	0.8722
46	0.7256	0.7496	0.7621	0.8409	0.8508	0.8682
51	0.7291	0.7518	0.7636	0.8385	0.8481	0.8648
61	0.7347	0.7554	0.7662	0.8349	0.8434	0.8592
71	0.7393	0.7583	0.7683	0.8321	0.8403	0.8549
81	0.7430	0.7607	0.7700	0.8298	0.8376	0.8515
91	0.7460	0.7626	0.7714	0.8279	0.8353	0.8484
101	0.7487	0.7644	0.7726	0.8264	0.8344	0.8460
201	0.7629	0.7738	0.7796	0.8178	0.8220	0.8322
301	0.7693	0.7781	0.7828	0.8140	0.8183	0.8260
401	0.7731	0.7807	0.7847	0.8118	0.8155	0.8223
501	0.7757	0.7825	0.7861	0.8103	0.8136	0.8198
601	0.7776	0.7838	0.7873	0.8092	0.8123	0.8179
701	0.7791	0.7848	0.7878	0.8084	0.8112	0.8164
801	0.7803	0.7857	0.7885	0.8077	0.8103	0.8152
901	0.7814	0.7864	0.7890	0.8071	0.8096	0.8142
1001	0.7822	0.7869	0.7894	0.8066	0.8090	0.8134

SOURCE: Pearson and Hartley, 1966.

Table 9. *Transformation of r to Z* $\left(i.e.\ Z = \frac{1}{2} \log e \frac{1+r}{1-r}\right)$.

r	Z	r	Z	r	Z
.00	.000				
.01	.010	.36	.377	.71	.887
.02	.020	.37	.388	.72	.908
.03	.030	.38	.400	.73	.929
.04	.040	.39	.412	.74	.950
.05	.050	.40	.424	.75	973
.06	.060	.41	.436	.76	.996
.07	.070	.42	.448	.77	1.020
.08	.080	.43	.460	.78	1.045
.09	.090	.44	.472	.79	1.071
.10	.100	.45	.485	.80	1.099
.11	.110	.46	.497	.81	1.127
.12	.121	.47	.510	.82	1.157
.13	.131	.48	.523	.83	1.188
.14	.141	.49	.536	.84	1.221
.15	.151	.50	.549	.85	1.256
.16	.161	.51	.563	.86	1.293
.17	.172	.52	.576	.87	1.333
.18	.182	.53	.590	.88	1.376
.19	.192	.54	.604	.89	1.422
.20	.203	.55	.618	.90	1.472
.21	.213	.56	.633	.91	1.528
.22	.224	.57	.648	.92	1.589
.23	.234	.58	.662	.93	1.658
.24	.245	.59	.678	.94	1.738
.25	.255	.60	.693	.95	1.832
.26	.266	.61	.709	.96	1.946
.27	.277	.62	.725	.97	2.092
.28	.288	.63	.741	.98	2.298
.29	.299	.64	.758	.99	2.647
.30	.310	.65	.775		
.31	.321	.66	.793		
.32	.332	.67	.811		
.33	.343	.68	.829		
.34	.354	.69	.848		
.35	.365	.70	.867		

Table 10.1. *95% Significance Points of the F-Distribution.*

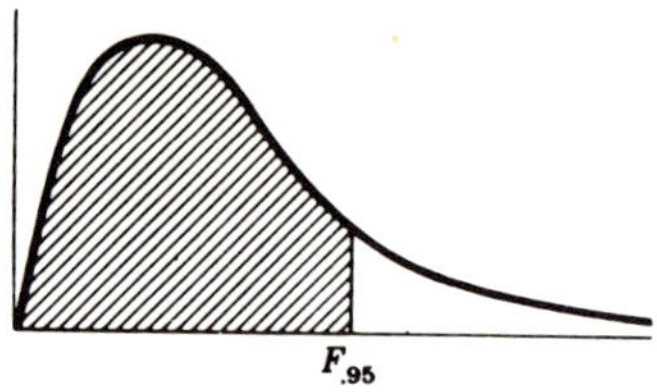

ν_2 \ ν_1	1	2	3	4	5	6	7	8	9
1	161·4	199·5	215·7	224·6	230·2	234·0	236·8	238·9	240·5
2	18·51	19·00	19·16	19·25	19·30	19·33	19·35	19·37	19·38
3	10·13	9·55	9·28	9·12	9·01	8·94	8·89	8·85	8·81
4	7·71	6·94	6·59	6·39	6·26	6·16	6·09	6·04	6·00
5	6·61	5·79	5·41	5·19	5·05	4·95	4·88	4·82	4·77
6	5·99	5·14	4·76	4·53	4·39	4·28	4·21	4·15	4·10
7	5·59	4·74	4·35	4·12	3·97	3·87	3·79	3·73	3·68
8	5·32	4·46	4·07	3·84	3·69	3·58	3·50	3·44	3·39
9	5·12	4·26	3·86	3·63	3·48	3·37	3·29	3·23	3·18
10	4·96	4·10	3·71	3·48	3·33	3·22	3·14	3·07	3·02
11	4·84	3·98	3·59	3·36	3·20	3·09	3·01	2·95	2·90
12	4·75	3·89	3·49	3·26	3·11	3·00	2·91	2·85	2·80
13	4·67	3·81	3·41	3·18	3·03	2·92	2·83	2·77	2·71
14	4·60	3·74	3·34	3·11	2·96	2·85	2·76	2·70	2·65
15	4·54	3·68	3·29	3·06	2·90	2·79	2·71	2·64	2·59
16	4·49	3·63	3·24	3·01	2·85	2·74	2·66	2·59	2·54
17	4·45	3·59	3·20	2·96	2·81	2·70	2·61	2·55	2·49
18	4·41	3·55	3·16	2·93	2·77	2·66	2·58	2·51	2·46
19	4·38	3·52	3·13	2·90	2·74	2·63	2·54	2·48	2·42
20	4·35	3·49	3·10	2·87	2·71	2·60	2·51	2·45	2·39
21	4·32	3·47	3·07	2·84	2·68	2·57	2·49	2·42	2·37
22	4·30	3·44	3·05	2·82	2·66	2·55	2·46	2·40	2·34
23	4·28	3·42	3·03	2·80	2·64	2·53	2·44	2·37	2·32
24	4·26	3·40	3·01	2·78	2·62	2·51	2·42	2·36	2·30
25	4·24	3·39	2·99	2·76	2·60	2·49	2·40	2·34	2·28
26	4·23	3·37	2·98	2·74	2·59	2·47	2·39	2·32	2·27
27	4·21	3·35	2·96	2·73	2·57	2·46	2·37	2·31	2·25
28	4·20	3·34	2·95	2·71	2·56	2·45	2·36	2·29	2·24
29	4·18	3·33	2·93	2·70	2·55	2·43	2·35	2·28	2·22
30	4·17	3·32	2·92	2·69	2·53	2·42	2·33	2·27	2·21
40	4·08	3·23	2·84	2·61	2·45	2·34	2·25	2·18	2·12
60	4·00	3·15	2·76	2·53	2·37	2·25	2·17	2·10	2·04
120	3·92	3·07	2·68	2·45	2·29	2·17	2·09	2·02	1·96
∞	3·84	3·00	2·60	2·37	2·21	2·10	2·01	1·94	1·88

SOURCE: Pearson and Hartley, 1966.

Table 10.1 *(continued)*

ν_2 \ ν_1	10	12	15	20	24	30	40	60	120	∞
1	241·9	243·9	245·9	248·0	249·1	250·1	251·1	252·2	253·3	254·3
2	19·40	19·41	19·43	19·45	19·45	19·46	19·47	19·48	19·49	19·50
3	8·79	8·74	8·70	8·66	8·64	8·62	8·59	8·57	8·55	8·53
4	5·96	5·91	5·86	5·80	5·77	5·75	5·72	5·69	5·66	5·63
5	4·74	4·68	4·62	4·56	4·53	4·50	4·46	4·43	4·40	4·36
6	4·06	4·00	3·94	3·87	3·84	3·81	3·77	3·74	3·70	3·67
7	3·64	3·57	3·51	3·44	3·41	3·38	3·34	3·30	3·27	3·23
8	3·35	3·28	3·22	3·15	3·12	3·08	3·04	3·01	2·97	2·93
9	3·14	3·07	3·01	2·94	2·90	2·86	2·83	2·79	2·75	2·71
10	2·98	2·91	2·85	2·77	2·74	2·70	2·66	2·62	2·58	2·54
11	2·85	2·79	2·72	2·65	2·61	2·57	2·53	2·49	2·45	2·40
12	2·75	2·69	2·62	2·54	2·51	2·47	2·43	2·38	2·34	2·30
13	2·67	2·60	2·53	2·46	2·42	2·38	2·34	2·30	2·25	2·21
14	2·60	2·53	2·46	2·39	2·35	2·31	2·27	2·22	2·18	2·13
15	2·54	2·48	2·40	2·33	2·29	2·25	2·20	2·16	2·11	2·07
16	2·49	2·42	2·35	2·28	2·24	2·19	2·15	2·11	2·06	2·01
17	2·45	2·38	2·31	2·23	2·19	2·15	2·10	2·06	2·01	1·96
18	2·41	2·34	2·27	2·19	2·15	2·11	2·06	2·02	1·97	1·92
19	2·38	2·31	2·23	2·16	2·11	2·07	2·03	1·98	1·93	1·88
20	2·35	2·28	2·20	2·12	2·08	2·04	1·99	1·95	1·90	1·84
21	2·32	2·25	2·18	2·10	2·05	2·01	1·96	1·92	1·87	1·81
22	2·30	2·23	2·15	2·07	2·03	1·98	1·94	1·89	1·84	1·78
23	2·27	2·20	2·13	2·05	2·01	1·96	1·91	1·86	1·81	1·76
24	2·25	2·18	2·11	2·03	1·98	1·94	1·89	1·84	1·79	1·73
25	2·24	2·16	2·09	2·01	1·96	1·92	1·87	1·82	1·77	1·71
26	2·22	2·15	2·07	1·99	1·95	1·90	1·85	1·80	1·75	1·69
27	2·20	2·13	2·06	1·97	1·93	1·88	1·84	1·79	1·73	1·67
28	2·19	2·12	2·04	1·96	1·91	1·87	1·82	1·77	1·71	1·65
29	2·18	2·10	2·03	1·94	1·90	1·85	1·81	1·75	1·70	1·64
30	2·16	2·09	2·01	1·93	1·89	1·84	1·79	1·74	1·68	1·62
40	2·08	2·00	1·92	1·84	1·79	1·74	1·69	1·64	1·58	1·51
60	1·99	1·92	1·84	1·75	1·70	1·65	1·59	1·53	1·47	1·39
120	1·91	1·83	1·75	1·66	1·61	1·55	1·50	1·43	1·35	1·25
∞	1·83	1·75	1·67	1·57	1·52	1·46	1·39	1·32	1·22	1·00

Table 10.2. *99% Significance Points of the F-Distribution.*

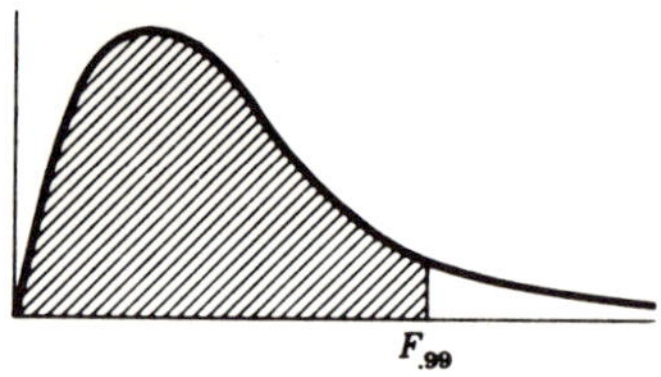

ν_2 \ ν_1	1	2	3	4	5	6	7	8	9
1	4052	4999·5	5403	5625	5764	5859	5928	5981	6022
2	98·50	99·00	99·17	99·25	99·30	99·33	99·36	99·37	99·39
3	34·12	30·82	29·46	28·71	28·24	27·91	27·67	27·49	27·35
4	21·20	18·00	16·69	15·98	15·52	15·21	14·98	14·80	14·66
5	16·26	13·27	12·06	11·39	10·97	10·67	10·46	10·29	10·16
6	13·75	10·92	9·78	9·15	8·75	8·47	8·26	8·10	7·98
7	12·25	9·55	8·45	7·85	7·46	7·19	6·99	6·84	6·72
8	11·26	8·65	7·59	7·01	6·63	6·37	6·18	6·03	5·91
9	10·56	8·02	6·99	6·42	6·06	5·80	5·61	5·47	5·35
10	10·04	7·56	6·55	5·99	5·64	5·39	5·20	5·06	4·94
11	9·65	7·21	6·22	5·67	5·32	5·07	4·89	4·74	4·63
12	9·33	6·93	5·95	5·41	5·06	4·82	4·64	4·50	4·39
13	9·07	6·70	5·74	5·21	4·86	4·62	4·44	4·30	4·19
14	8·86	6·51	5·56	5·04	4·69	4·46	4·28	4·14	4·03
15	8·68	6·36	5·42	4·89	4·56	4·32	4·14	4·00	3·89
16	8·53	6·23	5·29	4·77	4·44	4·20	4·03	3·89	3·78
17	8·40	6·11	5·18	4·67	4·34	4·10	3·93	3·79	3·68
18	8·29	6·01	5·09	4·58	4·25	4·01	3·84	3·71	3·60
19	8·18	5·93	5·01	4·50	4·17	3·94	3·77	3·63	3·52
20	8·10	5·85	4·94	4·43	4·10	3·87	3·70	3·56	3·46
21	8·02	5·78	4·87	4·37	4·04	3·81	3·64	3·51	3·40
22	7·95	5·72	4·82	4·31	3·99	3·76	3·59	3·45	3·35
23	7·88	5·66	4·76	4·26	3·94	3·71	3·54	3·41	3·30
24	7·82	5·61	4·72	4·22	3·90	3·67	3·50	3·36	3·26
25	7·77	5·57	4·68	4·18	3·85	3·63	3·46	3·32	3·22
26	7·72	5·53	4·64	4·14	3·82	3·59	3·42	3·29	3·18
27	7·68	5·49	4·60	4·11	3·78	3·56	3·39	3·26	3·15
28	7·64	5·45	4·57	4·07	3·75	3·53	3·36	3·23	3·12
29	7·60	5·42	4·54	4·04	3·73	3·50	3·33	3·20	3·09
30	7·56	5·39	4·51	4·02	3·70	3·47	3·30	3·17	3·07
40	7·31	5·18	4·31	3·83	3·51	3·29	3·12	2·99	2·89
60	7·08	4·98	4·13	3·65	3·34	3·12	2·95	2·82	2·72
120	6·85	4·79	3·95	3·48	3·17	2·96	2·79	2·66	2·56
∞	6·63	4·61	3·78	3·32	3·02	2·80	2·64	2·51	2·41

SOURCE: Pearson and Hartley, 1966.

Table 10.2 *(continued)*

ν_2 \ ν_1	10	12	15	20	24	30	40	60	120	∞
1	6056	6106	6157	6209	6235	6261	6287	6313	6339	6366
2	99·40	99·42	99·43	99·45	99·46	99·47	99·47	99·48	99·49	99·50
3	27·23	27·05	26·87	26·69	26·60	26·50	26·41	26·32	26·22	26·13
4	14·55	14·37	14·20	14·02	13·93	13·84	13·75	13·65	13·56	13·46
5	10·05	9·89	9·72	9·55	9·47	9·38	9·29	9·20	9·11	9·02
6	7·87	7·72	7·56	7·40	7·31	7·23	7·14	7·06	6·97	6·88
7	6·62	6·47	6·31	6·16	6·07	5·99	5·91	5·82	5·74	5·65
8	5·81	5·67	5·52	5·36	5·28	5·20	5·12	5·03	4·95	4·86
9	5·26	5·11	4·96	4·81	4·73	4·65	4·57	4·48	4·40	4·31
10	4·85	4·71	4·56	4·41	4·33	4·25	4·17	4·08	4·00	3·91
11	4·54	4·40	4·25	4·10	4·02	3·94	3·86	3·78	3·69	3·60
12	4·30	4·16	4·01	3·86	3·78	3·70	3·62	3·54	3·45	3·36
13	4·10	3·96	3·82	3·66	3·59	3·51	3·43	3·34	3·25	3·17
14	3·94	3·80	3·66	3·51	3·43	3·35	3·27	3·18	3·09	3·00
15	3·80	3·67	3·52	3·37	3·29	3·21	3·13	3·05	2·96	2·87
16	3·69	3·55	3·41	3·26	3·18	3·10	3·02	2·93	2·84	2·75
17	3·59	3·46	3·31	3·16	3·08	3·00	2·92	2·83	2·75	2·65
18	3·51	3·37	3·23	3·08	3·00	2·92	2·84	2·75	2·66	2·57
19	3·43	3·30	3·15	3·00	2·92	2·84	2·76	2·67	2·58	2·49
20	3·37	3·23	3·09	2·94	2·86	2·78	2·69	2·61	2·52	2·42
21	3·31	3·17	3·03	2·88	2·80	2·72	2·64	2·55	2·46	2·36
22	3·26	3·12	2·98	2·83	2·75	2·67	2·58	2·50	2·40	2·31
23	3·21	3·07	2·93	2·78	2·70	2·62	2·54	2·45	2·35	2·26
24	3·17	3·03	2·89	2·74	2·66	2·58	2·49	2·40	2·31	2·21
25	3·13	2·99	2·85	2·70	2·62	2·54	2·45	2·36	2·27	2·17
26	3·09	2·96	2·81	2·66	2·58	2·50	2·42	2·33	2·23	2·13
27	3·06	2·93	2·78	2·63	2·55	2·47	2·38	2·29	2·20	2·10
28	3·03	2·90	2·75	2·60	2·52	2·44	2·35	2·26	2·17	2·06
29	3·00	2·87	2·73	2·57	2·49	2·41	2·33	2·23	2·14	2·03
30	2·98	2·84	2·70	2·55	2·47	2·39	2·30	2·21	2·11	2·01
40	2·80	2·66	2·52	2·37	2·29	2·20	2·11	2·02	1·92	1·80
60	2·63	2·50	2·35	2·20	2·12	2·03	1·94	1·84	1·73	1·60
120	2·47	2·34	2·19	2·03	1·95	1·86	1·76	1·66	1·53	1·38
∞	2·32	2·18	2·04	1·88	1·79	1·70	1·59	1·47	1·32	1·00

Table 11. *Upper Percentage Points of Range for Students.*

Error *df*	α	*p* = number of									
		2	3	4	5	6	7	8	9	10	11
5	.05	3.64	4.60	5.22	5.67	6.03	6.33	6.58	6.80	6.99	7.17
	.01	5.70	6.97	7.80	8.42	8.91	9.32	9.67	9.97	10.24	10.48
6	.05	3.46	4.34	4.90	5.31	5.63	5.89	6.12	6.32	6.49	6.65
	.01	5.24	6.33	7.03	7.56	7.97	8.32	8.61	8.87	9.10	9.30
7	.05	3.34	4.16	4.68	5.06	5.36	5.61	5.82	6.00	6.16	6.30
	.01	4.95	5.92	6.54	7.01	7.37	7.68	7.94	8.17	8.37	8.55
8	.05	3.26	4.04	4.53	4.89	5.17	5.40	5.60	5.77	5.92	6.05
	.01	4.74	5.63	6.20	6.63	6.96	7.24	7.47	7.68	7.87	8.03
9	.05	3.20	3.95	4.42	4.76	5.02	5.24	5.43	5.60	5.74	5.87
	.01	4.60	5.43	5.96	6.35	6.66	6.91	7.13	7.32	7.49	7.65
10	.05	3.15	3.88	4.33	4.65	4.91	5.12	5.30	5.46	5.60	5.72
	.01	4.48	5.27	5.77	6.14	6.43	6.67	6.87	7.05	7.21	7.36
11	.05	3.11	3.82	4.26	4.57	4.82	5.03	5.20	5.35	5.49	5.61
	.01	4.39	5.14	5.62	5.97	6.25	6.48	6.67	6.84	6.99	7.13
12	.05	3.08	3.77	4.20	4.51	4.75	4.95	5.12	5.27	5.40	5.51
	.01	4.32	5.04	5.50	5.84	6.10	6.32	6.51	6.67	6.81	6.94
13	.05	3.06	3.73	4.15	4.45	4.69	4.88	5.05	5.19	5.32	5.43
	.01	4.26	4.96	5.40	5.73	5.98	6.19	6.37	6.53	6.67	6.79
14	.05	3.03	3.70	4.11	4.41	4.64	4.83	4.99	5.13	5.25	5.36
	.01	4.21	4.89	5.32	5.63	5.88	6.08	6.26	6.41	6.54	6.66
15	.05	3.01	3.67	4.08	4.37	4.60	4.78	4.94	5.08	5.20	5.31
	.01	4.17	4.83	5.25	5.56	5.80	5.99	6.16	6.31	6.44	6.55
16	.05	3.00	3.65	4.05	4.33	4.56	4.74	4.90	5.03	5.15	5.26
	.01	4.13	4.78	5.19	5.49	5.72	5.92	6.08	6.22	6.35	6.46
17	.05	2.98	3.63	4.02	4.30	4.52	4.71	4.86	4.99	5.11	5.21
	.01	4.10	4.74	5.14	5.43	5.66	5.85	6.01	6.15	6.27	6.38
18	.05	2.97	3.61	4.00	4.28	4.49	4.67	4.82	4.96	5.07	5.17
	.01	4.07	4.70	5.09	5.38	5.60	5.79	5.94	6.08	6.20	6.31
19	.05	2.96	3.59	3.98	4.25	4.47	4.65	4.79	4.92	5.04	5.14
	.01	4.05	4.67	5.05	5.33	5.55	5.73	5.89	6.02	6.14	6.25
20	.05	2.95	3.58	3.96	4.23	4.45	4.62	4.77	4.90	5.01	5.11
	.01	4.02	4.64	5.02	5.29	5.51	5.69	5.84	5.97	6.09	6.19
24	.05	2.92	3.53	3.90	4.17	4.37	4.54	4.68	4.81	4.92	5.01
	.01	3.96	4.54	4.91	5.17	5.37	5.54	5.69	5.81	5.92	6.02
30	.05	2.89	3.49	3.84	4.10	4.30	4.46	4.60	4.72	4.83	4.92
	.01	3.89	4.45	4.80	5.05	5.24	5.40	5.54	5.65	5.76	5.85
40	.05	2.86	3.44	3.79	4.04	4.23	4.39	4.52	4.63	4.74	4.82
	.01	3.82	4.37	4.70	4.93	5.11	5.27	5.39	5.50	5.60	5.69
60	.05	2.83	3.40	3.74	3.98	4.16	4.31	4.44	4.55	4.65	4.73
	.01	3.76	4.28	4.60	4.82	4.99	5.13	5.25	5.36	5.45	5.53
120	.05	2.80	3.36	3.69	3.92	4.10	4.24	4.36	4.48	4.56	4.64
	.01	3.70	4.20	4.50	4.71	4.87	5.01	5.12	5.21	5.30	5.38
∞	.05	2.77	3.31	3.63	3.86	4.03	4.17	4.29	4.39	4.47	4.55
	.01	3.64	4.12	4.40	4.60	4.76	4.88	4.99	5.08	5.16	5.23

SOURCE: Pearson and Hartley, 1966.

NOTE: $q_x = \dfrac{\bar{x}\text{max} - \bar{x}\text{min}}{s_{\bar{x}}}$

treatment means									α	Error *df*
12	13	14	15	16	17	18	19	20		
7.32	7.47	7.60	7.72	7.83	7.93	8.03	8.12	8.21	.05	5
10.70	10.89	11.08	11.24	11.40	11.55	11.68	11.81	11.93	.01	
6.79	6.92	7.03	7.14	7.24	7.34	7.43	7.51	7.59	.05	6
9.49	9.65	9.81	9.95	10.08	10.21	10.32	10.43	10.54	.01	
6.43	6.55	6.66	6.76	6.85	6.94	7.02	7.09	7.17	.05	7
8.71	8.86	9.00	9.12	9.24	9.35	9.46	9.55	9.65	.01	
6.18	6.29	6.39	6.48	6.57	6.65	6.73	6.80	6.87	.05	8
8.18	8.31	8.44	8.55	8.66	8.76	8.85	8.94	9.03	.01	
5.98	6.09	6.19	6.28	6.36	6.44	6.51	6.58	6.64	.05	9
7.78	7.91	8.03	8.13	8.23	8.32	8.41	8.49	8.57	.01	
5.83	5.93	6.03	6.11	6.20	6.27	6.34	6.40	6.47	.05	10
7.48	7.60	7.71	7.81	7.91	7.99	8.07	8.15	8.22	.01	
5.71	5.81	5.90	5.99	6.06	6.14	6.20	6.26	6.33	.05	11
7.25	7.36	7.46	7.56	7.65	7.73	7.81	7.88	7.95	.01	
5.62	5.71	5.80	5.88	5.95	6.03	6.09	6.15	6.21	.05	12
7.06	7.17	7.26	7.36	7.44	7.52	7.59	7.66	7.73	.01	
5.53	5.63	5.71	5.79	5.86	5.93	6.00	6.05	6.11	.05	13
6.90	7.01	7.10	7.19	7.27	7.34	7.42	7.48	7.55	.01	
5.46	5.55	5.64	5.72	5.79	5.85	5.92	5.97	6.03	.05	14
6.77	6.87	6.96	7.05	7.12	7.20	7.27	7.33	7.39	.01	
5.40	5.49	5.58	5.65	5.72	5.79	5.85	5.90	5.96	.05	15
6.66	6.76	6.84	6.93	7.00	7.07	7.14	7.20	7.26	.01	
5.35	5.44	5.52	5.59	5.66	5.72	5.79	5.84	5.90	.05	16
6.56	6.66	6.74	6.82	6.90	6.97	7.03	7.09	7.15	.01	
5.31	5.39	5.47	5.55	5.61	5.68	5.74	5.79	5.84	.05	17
6.48	6.57	6.66	6.73	6.80	6.87	6.94	7.00	7.05	.01	
5.27	5.35	5.43	5.50	5.57	5.63	5.69	5.74	5.79	.05	18
6.41	6.50	6.58	6.65	6.72	6.79	6.85	6.91	6.96	.01	
5.23	5.32	5.39	5.46	5.53	5.59	5.65	5.70	5.75	.05	19
6.34	6.43	6.51	6.58	6.65	6.72	6.78	6.84	6.89	.01	
5.20	5.28	5.36	5.43	5.49	5.55	5.61	5.66	5.71	.05	20
6.29	6.37	6.45	6.52	6.59	6.65	6.71	6.76	6.82	.01	
5.10	5.18	5.25	5.32	5.38	5.44	5.50	5.54	5.59	.05	24
6.11	6.19	6.26	6.33	6.39	6.45	6.51	6.56	6.61	.01	
5.00	5.08	5.15	5.21	5.27	5.33	5.38	5.43	5.48	.05	30
5.93	6.01	6.08	6.14	6.20	6.26	6.31	6.36	6.41	.01	
4.91	4.98	5.05	5.11	5.16	5.22	5.27	5.31	5.36	.05	40
5.77	5.84	5.90	5.96	6.02	6.07	6.12	6.17	6.21	.01	
4.81	4.88	4.94	5.00	5.06	5.11	5.16	5.20	5.24	.05	60
5.60	5.67	5.73	5.79	5.84	5.89	5.93	5.98	6.02	.01	
4.72	4.78	4.84	4.90	4.95	5.00	5.05	5.09	5.13	.05	120
5.44	5.51	5.56	5.61	5.66	5.71	5.75	5.79	5.83	.01	
4.62	4.68	4.74	4.80	4.85	4.89	4.93	4.97	5.01	.05	∞
5.29	5.35	5.40	5.45	5.49	5.54	5.57	5.61	5.65	.01	

Table 12. *Significant Ranges for Students of 5% and 1% Level New Multiple Range Test.*

Error df	Protection level	p = number of means for range being tested													
		2	3	4	5	6	7	8	9	10	12	14	16	18	20
1	.05	18.0	18.0	18.0	18.0	18.0	18.0	18.0	18.0	18.0	18.0	18.0	18.0	18.0	18.0
	.01	90.0	90.0	90.0	90.0	90.0	90.0	90.0	90.0	90.0	90.0	90.0	90.0	90.0	90.0
2	.05	6.09	6.09	6.09	6.09	6.09	6.09	6.09	6.09	6.09	6.09	6.09	6.09	6.09	6.09
	.01	14.0	14.0	14.0	14.0	14.0	14.0	14.0	14.0	14.0	14.0	14.0	14.0	14.0	14.0
3	.05	4.50	4.50	4.50	4.50	4.50	4.50	4.50	4.50	4.50	4.50	4.50	4.50	4.50	4.50
	.01	8.26	8.5	8.6	8.7	8.8	8.9	8.9	9.0	9.0	9.0	9.1	9.2	9.3	9.3
4	.05	3.93	4.01	4.02	4.02	4.02	4.02	4.02	4.02	4.02	4.02	4.02	4.02	4.02	4.02
	.01	6.51	6.8	6.9	7.0	7.1	7.1	7.2	7.2	7.3	7.3	7.4	7.4	7.5	7.5
5	.05	3.64	3.74	3.79	3.83	3.83	3.83	3.83	3.83	3.83	3.83	3.83	3.83	3.83	3.83
	.01	5.70	5.96	6.11	6.18	6.26	6.33	6.40	6.44	6.5	6.6	6.6	6.7	6.7	6.8
6	.05	3.46	3.58	3.64	3.68	3.68	3.68	3.68	3.68	3.68	3.68	3.68	3.68	3.68	3.68
	.01	5.24	5.51	5.65	5.73	5.81	5.88	5.95	6.00	6.0	6.1	6.2	6.2	6.3	6.3
7	.05	3.35	3.47	3.54	3.58	3.60	3.61	3.61	3.61	3.61	3.61	3.61	3.61	3.61	3.61
	.01	4.95	5.22	5.37	5.45	5.53	5.61	5.69	5.73	5.8	5.8	5.9	5.9	6.0	6.0
8	.05	3.26	3.39	3.47	3.52	3.55	3.56	3.56	3.56	3.56	3.56	3.56	3.56	3.56	3.56
	.01	4.74	5.00	5.14	5.23	5.32	5.40	5.47	5.51	5.5	5.6	5.7	5.7	5.8	5.8
9	.05	3.20	3.34	3.41	3.47	3.50	3.52	3.52	3.52	3.52	3.52	3.52	3.52	3.52	3.52
	.01	4.60	4.86	4.99	5.08	5.17	5.25	5.32	5.36	5.4	5.5	5.5	5.6	5.7	5.7
10	.05	3.15	3.30	3.37	3.43	3.46	3.47	3.47	3.47	3.47	3.47	3.47	3.47	3.47	3.48
	.01	4.48	4.73	4.88	4.96	5.06	5.13	5.20	5.24	5.28	5.36	5.42	5.48	5.54	5.55
11	.05	3.11	3.27	3.35	3.39	3.43	3.44	3.45	3.46	4.46	3.46	3.46	3.46	3.47	3.48
	.01	4.39	4.63	4.77	4.86	4.94	5.01	5.06	5.12	5.15	5.24	5.28	5.34	5.38	5.39
12	.05	3.08	3.23	3.33	3.36	3.40	3.42	3.44	3.44	3.46	3.46	3.46	3.46	3.47	3.48
	.01	4.32	4.55	4.68	4.76	4.84	4.92	4.96	5.02	5.07	5.13	5.17	5.22	5.24	5.26
13	.05	3.06	3.21	3.30	3.35	3.38	3.41	3.42	3.44	3.45	3.45	3.46	3.46	3.47	3.47
	.01	4.26	4.48	4.62	4.69	4.74	4.84	4.88	4.94	4.98	5.04	5.08	5.13	5.14	5.15
14	.05	3.03	3.18	3.27	3.33	3.37	3.39	3.41	3.42	3.44	3.45	3.46	3.46	3.47	3.47
	.01	4.21	4.42	4.55	4.63	4.70	4.78	4.83	4.87	4.91	4.96	5.00	5.04	5.06	5.07
15	.05	3.01	3.16	3.25	3.31	3.36	3.38	3.40	3.42	3.43	3.44	3.45	3.46	3.47	3.47
	.01	4.17	4.37	4.50	4.58	4.64	4.72	4.77	4.81	4.84	4.90	4.94	4.97	4.99	5.00

Error *df*	Protection level	*p* = number of means for range being tested													
		2	3	4	5	6	7	8	9	10	12	14	16	18	20
16	.05	3.00	3.15	3.23	3.30	3.34	3.37	3.39	3.41	3.43	3.44	3.45	3.46	3.47	3.47
	.01	4.13	4.34	4.45	4.54	4.60	4.67	4.72	4.76	4.79	4.84	4.88	4.91	4.93	4.94
17	.05	2.98	3.13	3.22	3.28	3.33	3.36	3.38	3.40	3.42	3.44	3.45	3.46	3.47	3.47
	.01	4.10	4.30	4.41	4.50	4.56	4.63	4.68	4.72	4.75	4.80	4.83	4.86	4.88	4.89
18	.05	2.97	3.12	3.21	3.27	3.32	3.35	3.37	3.39	3.41	3.43	3.45	3.46	3.47	3.47
	.01	4.07	4.27	4.38	4.46	4.53	4.59	4.64	4.68	4.71	4.76	4.79	4.82	4.84	4.85
19	.05	2.96	3.11	3.19	3.26	3.31	3.35	3.37	3.39	3.41	3.43	3.44	3.46	3.47	3.47
	.01	4.05	4.24	4.35	4.43	4.50	4.56	4.61	4.64	4.67	4.72	4.76	4.79	4.81	4.82
20	.05	2.95	3.10	3.18	3.25	3.30	3.34	3.36	3.38	3.40	3.43	3.44	3.46	3.46	3.47
	.01	4.02	4.22	4.33	4.40	4.47	4.53	4.58	4.61	4.65	4.69	4.73	4.76	4.78	4.79
22	.05	2.93	3.08	3.17	3.24	3.29	3.32	3.35	3.37	3.39	3.42	3.44	3.45	3.46	3.47
	.01	3.99	4.17	4.28	4.36	4.42	4.48	4.53	4.57	4.60	4.65	4.68	4.71	4.74	4.75
24	.05	2.92	3.07	3.15	3.22	3.28	3.31	3.34	3.37	3.38	3.41	3.44	3.45	3.46	3.47
	.01	3.96	4.14	4.24	4.33	4.39	4.44	4.49	4.53	4.57	4.62	4.64	4.67	4.70	4.72
26	.05	2.91	3.06	3.14	3.21	3.27	3.30	3.34	3.36	3.38	3.41	3.43	3.45	3.46	3.47
	.01	3.93	4.11	4.21	4.30	4.36	4.41	4.46	4.50	4.53	4.58	4.62	4.65	4.67	4.69
28	.05	2.90	3.04	3.13	3.20	3.26	3.30	3.33	3.35	3.37	3.40	3.43	3.45	3.46	3.47
	.01	3.91	4.08	4.18	4.28	4.34	4.39	4.43	4.47	4.51	4.56	4.60	4.62	4.65	4.67
30	.05	2.89	3.04	3.12	3.20	3.25	3.29	3.32	3.35	3.37	3.40	3.43	3.44	3.46	3.47
	.01	3.89	4.06	4.16	4.22	4.32	4.36	4.41	4.45	4.48	4.54	4.58	4.61	4.63	4.65
40	.05	2.86	3.01	3.10	3.17	3.22	3.27	3.30	3.33	3.35	3.39	3.42	3.44	3.46	3.47
	.01	3.82	3.99	4.10	4.17	4.24	4.30	4.34	4.37	4.41	4.46	4.51	4.54	4.57	4.59
60	.05	2.83	2.98	3.08	3.14	3.20	3.24	3.28	3.31	3.33	3.37	3.40	3.43	3.45	3.47
	.01	3.76	3.92	4.03	4.12	4.17	4.23	4.27	4.31	4.34	4.39	4.44	4.47	4.50	4.53
100	.05	2.80	2.95	3.05	3.12	3.18	3.22	3.26	3.29	3.32	3.36	3.40	3.42	3.45	3.47
	.01	3.71	3.86	3.98	4.06	4.11	4.17	4.21	4.25	4.29	4.35	4.38	4.42	4.45	4.48
∞	.05	2.77	2.92	3.02	3.09	3.15	3.19	3.23	3.26	3.29	3.34	3.38	3.41	3.44	3.47
	.01	3.64	3.80	3.90	3.98	4.04	4.09	4.14	4.17	4.20	4.26	4.31	4.34	4.38	4.41

SOURCE: Reprinted from D. B. Duncan, *Multiple Range and Multiple F Tests.* BIOMETRICS 11: 1–42, 1955, with permission of the Biometric Society and the author.

Table 13. *Percentage Points of the Ratio,* s^2_{max}/s^2_{min}.

Upper 5% *points*

ν \ k	2	3	4	5	6	7	8	9	10	11	12
2	39·0	87·5	142	202	266	333	403	475	550	626	704
3	15·4	27·8	39·2	50·7	62·0	72·9	83·5	93·9	104	114	124
4	9·60	15·5	20·6	25·2	29·5	33·6	37·5	41·1	44·6	48·0	51·4
5	7·15	10·8	13·7	16·3	18·7	20·8	22·9	24·7	26·5	28·2	29·9
6	5·82	8·38	10·4	12·1	13·7	15·0	16·3	17·5	18·6	19·7	20·7
7	4·99	6·94	8·44	9·70	10·8	11·8	12·7	13·5	14·3	15·1	15·8
8	4·43	6·00	7·18	8·12	9·03	9·78	10·5	11·1	11·7	12·2	12·7
9	4·03	5·34	6·31	7·11	7·80	8·41	8·95	9·45	9·91	10·3	10·7
10	3·72	4·85	5·67	6·34	6·92	7·42	7·87	8·28	8·66	9·01	9·34
12	3·28	4·16	4·79	5·30	5·72	6·09	6·42	6·72	7·00	7·25	7·48
15	2·86	3·54	4·01	4·37	4·68	4·95	5·19	5·40	5·59	5·77	5·93
20	2·46	2·95	3·29	3·54	3·76	3·94	4·10	4·24	4·37	4·49	4·59
30	2·07	2·40	2·61	2·78	2·91	3·02	3·12	3·21	3·29	3·36	3·39
60	1·67	1·85	1·96	2·04	2·11	2·17	2·22	2·26	2·30	2·33	2·36
∞	1·00	1·00	1·00	1·00	1·00	1·00	1·00	1·00	1·00	1·00	1·00

Upper 1% points

ν \ k	2	3	4	5	6	7	8	9	10	11	12
2	199	448	729	1036	1362	1705	2063	2432	2813	3204	3605
3	47·5	85	120	151	184	21(6)	24(9)	28(1)	31(0)	33(7)	36(1)
4	23·2	37	49	59	69	79	89	97	106	113	120
5	14·9	22	28	33	38	42	46	50	54	57	60
6	11·1	15·5	19·1	22	25	27	30	32	34	36	37
7	8·89	12·1	14·5	16·5	18·4	20	22	23	24	26	27
8	7·50	9·9	11·7	13·2	14·5	15·8	16·9	17·9	18·9	19·8	21
9	6·54	8·5	9·9	11·1	12·1	13·1	13·9	14·7	15·3	16·0	16·6
10	5·85	7·4	8·6	9·6	10·4	11·1	11·8	12·4	12·9	13·4	13·9
12	4·91	6·1	6·9	7·6	8·2	8·7	9·1	9·5	9·9	10·2	10·6
15	4·07	4·9	5·5	6·0	6·4	6·7	7·1	7·3	7·5	7·8	8·0
20	3·32	3·8	4·3	4·6	4·9	5·1	5·3	5·5	5·6	5·8	5·9
30	2·63	3·0	3·3	3·4	3·6	3·7	3·8	3·9	4·0	4·1	4·2
60	1·96	2·2	2·3	2·4	2·4	2·5	2·5	2·6	2·6	2·7	2·7
∞	1·00	1·0	1·0	1·0	1·0	1·0	1·0	1·0	1·0	1·0	1·0

SOURCE: Pearson and Hartley, 1966.

NOTE: s^2_{max} is the largest and s^2_{min} the smallest in a set of k independent mean squares, each based on υ degrees of freedom.

Values in the column $k = 2$ and in the rows $\upsilon = 2$ and ∞ are exact. Elsewhere the third digit may be in error by a few units for the 5% points and several units for the 1% points. The third digit figures in brackets for $\upsilon = 3$ are the most uncertain.

Table 14. *Factorials and Their Common Logarithms.*

No.	Factorial.	Logarithm.	No.	Factorial.	Logarithm.
1	1	0·000 0000	26	4·03291	26·605 6190
2	2	0·301 0300	27	1·08889	28·036 9828
3	6	0·778 1513	28	3·04888	29·484 1408
4	24	1·380 2112	29	8·84176	30·946 5388
5	120	2·079 1812	30	2·65253	32·423 6601
6	720	2·857 3325	31	8·22284	33·915 0218
7	5040	3·702 4305	32	2·63131	35·420 1717
8	40320	4·605 5205	33	8·68332	36·938 6857
9	362880	5·559 7630	34	2·95233	38·470 1646
10	3·62880	6·559 7630	35	1·03331	40·014 2326
11	3·99168	7·601 1557	36	3·71993	41·570 5351
12	4·79002	8·680 3370	37	1·37638	43·138 7369
13	6·22702	9·794 2803	38	5·23023	44·718 5205
14	8·71783	10·940 4084	39	2·03979	46·309 5851
15	1·30767	12·116 4996	40	8·15915	47·911 6451
16	2·09228	13·320 6196	41	3·34525	49·524 4289
17	3·55687	14·551 0685	42	1·40501	51·147 6782
18	6·40237	15·806 3410	43	6·04153	52·781 1467
19	1·21645	17·085 0946	44	2·65827	54·424 5993
20	2·43290	18·386 1246	45	1·19622	56·077 8119
21	5·10909	19·708 3439	46	5·50262	57·740 5697
22	1·12400	21·050 7666	47	2·58623	59·412 6676
23	2·58520	22·412 4944	48	1·24139	61·093 9088
24	6·20448	23·792 7057	49	6·08282	62·784 1049
25	1·55112	25·190 6457	50	3·04141	64·483 0749

NOTE: The power of 10 by which to multiply the factorial is given by the whole number of the logarithm.

Table 15. *Critical Values for the Wilcoxon Test for Paired Data.*

n	$2\alpha \leq 0.10$	$2\alpha \leq 0.05$	$2\alpha \leq 0.02$	$2\alpha \leq 0.01$
5	0- 15	-	-	-
6	2- 19	0- 21	-	-
7	3- 25	2- 26	0- 28	-
8	5- 31	3- 33	1- 35	0- 36
9	8- 37	5- 40	3- 42	1- 44
10	10- 45	8- 47	5- 50	3- 52
11	13- 53	10- 56	7- 59	5- 61
12	17- 61	13- 65	9- 69	7- 71
13	21- 70	17- 74	12- 79	9- 82
14	25- 80	21- 84	15- 90	12- 93
15	30- 90	25- 95	19-101	15-105
16	35-101	29-107	23-113	19-117
17	41-112	34-119	28-125	23-130
18	47-124	40-131	32-139	27-144
19	53-137	46-144	37-153	32-158
20	60-150	52-158	43-167	37-173
21	67-164	58-173	49-182	42-189
22	75-178	66-187	55-198	48-205
23	83-193	73-203	62-214	54-222
24	91-209	81-219	69-231	61-239
25	100-225	89-236	76-249	68-257

SOURCE: Reproduced from *Documenta Geigy Scientific Tables,* 7th edition, by permission of CIBA-GEIGY Limited, Basle, Switzerland.

Table 16. *Critical Points of Rank Sums (Two-tailed Alternatives).*

n_2 = larger n	P	n_1 = smaller n													
		2	3	4	5	6	7	8	9	10	11	12	13	14	15
4	.05			10											
	.01			—											
5	.05		6	11	17										
	.01		—	—	15										
6	.05		7	12	18	26									
	.01		—	10	16	23									
7	.05		7	13	20	27	36								
	.01		—	10	17	24	32								
8	.05	3	8	14	21	29	38	49							
	.01	—	—	11	17	25	34	43							
9	.05	3	8	15	22	31	40	51	63						
	.01	—	6	11	18	26	35	45	56						
10	.05	3	9	15	23	32	42	53	65	78					
	.01	—	6	12	19	27	37	47	58	71					
11	.05	4	9	16	24	34	44	55	68	81	96				
	.01	—	6	12	20	28	38	49	61	74	87				
12	.05	4	10	17	26	35	46	58	71	85	99	115			
	.01	—	7	13	21	30	40	51	63	76	90	106			
13	.05	4	10	18	27	37	48	60	73	88	103	119	137		
	.01	—	7	14	22	31	41	53	65	79	93	109	125		
14	.05	4	11	19	28	38	50	63	76	91	106	123	141	160	
	.01	—	7	14	22	32	43	54	67	81	96	112	129	147	
15	.05	4	11	20	29	40	52	65	79	94	110	127	145	164	185
	.01	—	8	15	23	33	44	56	70	84	99	115	133	151	171
16	.05	4	12	21	31	42	54	67	82	97	114	131	150	169	
	.01	—	8	15	24	34	46	58	72	86	102	119	137	155	
17	.05	5	12	21	32	43	56	70	84	100	117	135	154		
	.01	—	8	16	25	36	47	60	74	89	105	122	140		
18	.05	5	13	22	33	45	58	72	87	103	121	139			
	.01	—	8	16	26	37	49	62	76	92	108	125			
19	.05	5	13	23	34	46	60	74	90	107	124				
	.01	3	9	17	27	38	50	64	78	94	111				
20	.05	5	14	24	35	48	62	77	93	110					
	.01	3	9	18	28	39	52	66	81	97					
21	.05	6	14	25	37	50	64	79	95						
	.01	3	9	18	29	40	53	68	83						
22	.05	6	15	26	38	51	66	82							
	.01	3	10	19	29	42	55	70							
23	.05	6	15	27	39	53	68								
	.01	3	10	19	30	43	57								
24	.05	6	16	28	40	55									
	.01	3	10	20	31	44									
25	.05	6	16	28	42										
	.01	3	11	20	32										
26	.05	7	17	29											
	.01	3	11	21											
27	.05	7	17												
	.01	4	11												
28	.05	7													
	.01	4													

SOURCE: Reprinted from C. White, *The Use of Ranks in a Test of significance for comparing Two Treatments.* BIOMETRICS 8: 33–41 1952, with permission of the Biometric Society and the author.

Table 17.1. *Probabilities Associated with Obtaining Values as Large as Observed Values with Friedman's Test (k = 3).*

N = 2		N = 3		N = 4		N = 5	
χ_r^2	*p*	χ_r^2	*p*	χ_r^2	*p*	χ_r^2	*p*
0	1.000	.000	1.000	.0	1.000	.0	1.000
1	.833	.667	.944	.5	.931	.4	.954
3	.500	2.000	.528	1.5	.653	1.2	.691
4	.167	2.667	.361	2.0	.431	1.6	.522
		4.667	.194	3.5	.273	2.8	.367
		6.000	.028	4.5	.125	3.6	.182
				6.0	.069	4.8	.124
				6.5	.042	5.2	.093
				8.0	.0046	6.4	.039
						7.6	.024
						8.4	.0085
						10.0	.00077

N = 6		N = 7		N = 8		N = 9	
χ_r^2	*p*	χ_r^2	*p*	χ_r^2	*p*	χ_r^2	*p*
.00	1.000	.000	1.000	.00	1.000	.000	1.000
.33	.956	.286	.964	.25	.967	.222	.971
1.00	.740	.857	.768	.75	.794	.667	.814
1.33	.570	1.143	.620	1.00	.654	.889	.865
2.33	.430	2.000	.486	1.75	.531	1.556	.569
3.00	.252	2.571	.305	2.25	.355	2.000	.398
4.00	.184	3.429	.237	3.00	.285	2.667	.328
4.33	.142	3.714	.192	3.25	.236	2.889	.278
5.33	.072	4.571	.112	4.00	.149	3.556	.187
6.33	.052	5.429	.085	4.75	.120	4.222	.154
7.00	.029	6.000	.052	5.25	.079	4.667	.107
8.33	.012	7.143	.027	6.25	.047	5.556	.069
9.00	.0081	7.714	.021	6.75	.038	6.000	.057
9.33	.0055	8.000	.016	7.00	.038	6.222	.048
10.33	.0017	8.857	.0084	7.75	.018	6.889	.031
12.00	.00013	10.286	.0036	9.00	.0099	8.000	.019
		10.571	.0027	9.25	.0080	8.222	.016
		11.143	.0012	9.75	.0048	8.667	.010
		12.286	.00032	10.75	.0024	9.556	.0060
		14.000	.000021	12.00	.0011	10.667	.0035
				12.25	.00086	10.889	.0029
				13.00	.00026	11.556	.0013
				14.25	.000061	12.667	.00066
				16.00	.0000036	13.556	.00035
						14.000	.00020
						14.222	.000097
						14.889	.000054
						16.222	.000011
						18.000	.0000006

SOURCE: Adapted from Friedman, M. *The Use of Ranks to Avoid the Assumption of Normality Implicit in the Analysis of Variance*, J. Amer. Statist. Ass., 32, 688–689, 1937, with permission of the author and publisher.

Table 17.2. *Probabilities Associated with Obtaining Values as Large as Observed Values with Friedman's Test (k = 4).*

N = 2		N = 3		N = 4			
χ_r^2	p	χ_r^2	p	χ_r^2	p	χ_r^2	p
.0	1.000	.2	1.000	.0	1.000	5.7	.141
.6	.958	.6	.958	.3	.992	6.0	.105
1.2	.834	1.0	.910	.6	.928	6.3	.094
1.8	.792	1.8	.727	.9	.900	6.6	.077
2.4	.625	2.2	.608	1.2	.800	6.9	068
3.0	.542	2.6	.524	1.5	.754	7.2	.054
3.6	.458	3.4	.446	1.8	.677	7.5	.052
4.2	.375	3.8	.342	2.1	.649	7.8	.036
4.8	.208	4.2	.300	2.4	.524	8.1	.033
5.4	.167	5.0	.207	2.7	.508	8.4	.019
6.0	.042	5.4	.175	3.0	.432	8.7	.014
		5.8	.148	3.3	.389	9.3	.012
		6.6	.075	3.6	.355	9.6	.0069
		7.0	054	3.9	.324	9.9	.0062
		7.4	.033	4.5	.242	10.2	.0027
		8.2	.017	4.8	.200	10.8	.0016
		9.0	.0017	5.1	.190	11.1	.00094
				5.4	.158	12.0	.000072

SOURCE: Adapted from Friedman, M. *The Use of Ranks to Avoid the Assumption of Normality Implicit in the Analysis of Variance,* J. Amer. Statist. Ass., 32, 688–689, 1937, with permission of the author and the publisher.

Table 18. *Critical Values for a 5% Two-Sided Runs Test.*

	5	6	7	8	9	10	11	12	13	14	15	16	17	18	19	20
2								2 6	2 6	2 6	2 6	2 6	2 6	2 6	2 6	2 6
3		2 8	2 8	2 8	2 8	2 8	2 8	2 8	2 8	2 8	3 8	3 8	3 8	3 8	3 8	3 8
4	2 9	2 9	2 10	3 10	3 10	3 10	3 10	3 10	3 10	3 10	3 10	4 10	4 10	4 10	4 10	4 10
5	2 10	3 10	3 11	3 11	3 12	3 12	4 12	4 12	4 12	4 12	4 12	4 12	4 12	5 12	5 12	5 12
6		3 11	3 12	3 12	4 13	4 13	4 13	4 13	5 14	5 14	5 14	5 14	5 14	5 14	6 14	6 14
7			3 13	4 13	4 14	5 14	5 14	5 14	5 15	5 15	6 15	6 16	6 16	6 16	6 16	6 16
8				4 14	5 14	5 15	5 15	6 16	6 16	6 16	6 16	6 17	7 17	7 17	7 17	7 17
9					5 15	5 16	6 16	6 16	6 17	7 17	7 18	7 18	7 18	8 18	8 18	8 18
10						6 16	6 17	7 17	7 18	7 18	7 18	8 19	8 19	8 19	8 20	9 20
11							7 17	7 18	7 19	8 19	8 19	8 20	9 20	9 20	9 21	9 21
12								7 19	8 19	8 20	8 20	9 21	9 21	9 21	10 22	10 22
13									8 20	9 20	9 21	9 21	10 22	10 22	10 23	10 23
14										9 21	9 22	10 22	10 23	10 23	11 23	11 24
15											10 22	10 23	11 23	11 24	11 24	12 25
16												11 23	11 24	11 25	12 25	12 25
17													11 25	12 25	12 26	13 26
18														12 26	13 26	13 27
19															13 27	13 27
20																14 28

SOURCE: From C. Eisenhart and F. Swed, *Tables for Testing Randomness of Grouping in a Sequence of Alternatives,* Annals of Mathematical Statistics, Vol. 14, 1943, p. 66. Reprinted by permission of the authors and the publisher.

Table 19. *Random Numbers.*

93108	77033	68325	10160	38667	62441	87023	94372	06164	30700
28271	08589	83279	48838	60935	70541	53814	95588	05832	80235
21841	35545	11148	34775	17308	88034	97765	35959	52843	44895
22025	79554	19698	25255	50283	94037	57463	92925	12042	91414
09210	20779	02994	02258	86978	85092	54052	18354	20914	28460
90552	71129	03621	20517	16908	06668	29916	51537	93658	29525
01130	06995	20258	10351	99248	51660	38861	49668	74742	47181
22604	56719	21784	68788	38358	59827	19270	99287	81193	43366
06690	01800	34272	65497	94891	14537	91358	21587	95765	72605
59809	69982	71809	64984	48709	43991	24987	69246	86400	29559
56475	02726	58511	95405	70293	84971	06676	44075	32338	31980
02730	34870	83209	03138	07715	31557	55242	61308	26507	06186
74482	33990	13509	92588	10462	76546	46097	01825	20153	36271
19793	22487	94238	81054	95488	23617	15539	94335	73822	93481
19020	27856	60526	24144	98021	60564	46373	86928	52135	74919
69565	60635	65709	77887	42766	86698	14004	94577	27936	47220
69274	23208	61035	84263	15034	28717	76146	22021	23779	98562
83658	14204	09445	41081	49630	34215	89806	40930	97194	21747
78612	51102	66826	40430	54072	62164	68977	95583	11765	81072
14980	74158	78216	38985	60838	82836	42777	85321	90463	11813
63172	28010	29405	91554	75195	51183	65805	87525	35952	83204
71167	37984	52737	06869	38122	95322	41356	19391	96787	64410
78530	56410	19195	34434	83712	50397	80920	15464	81350	18673
98324	03774	07573	67864	06497	20758	83454	22756	83959	96347
55793	30055	08373	32652	02654	75980	02095	87545	88815	80086
05674	34471	61967	91266	38814	44728	32455	17057	08339	93997
15643	22245	07592	22078	73628	60902	41561	54608	41023	98345
66750	19609	70358	03622	64898	82220	69304	46235	97332	64539
42320	74314	50222	82339	51564	42885	50482	98501	02245	88990
73752	73818	15470	04914	24936	65514	56633	72030	30856	85183
97546	02188	46373	21486	28221	08155	23486	66134	88799	49496
32569	52162	38444	42004	78011	16909	94194	79732	47114	23919
36048	93973	82596	28739	86985	58144	65007	08786	14826	04896
40455	36702	38965	56042	80023	28169	04174	65533	52718	55255
33597	47071	55618	51796	71027	46690	08002	45066	02870	60012
22828	96380	35883	15910	17211	42358	14056	55438	98148	35384
00631	95925	19324	31497	88118	06283	84596	72091	53987	01477
75722	36478	07634	63114	27164	15467	03983	09141	60562	65725
80577	01771	61510	17099	28731	41426	18853	41523	14914	76661
10524	20900	65463	83680	05005	11611	64426	59065	06758	02892
93815	69446	75253	51915	97839	75427	90685	60352	96288	34248
81867	97119	93446	20862	46591	97677	42704	13718	44975	67145
64649	07689	16711	12169	15238	74106	60655	56289	74166	78561
55768	09210	52439	33355	57884	36791	00853	49969	74814	09270
38080	49460	48137	61589	42742	92035	21766	19435	92579	27683
22360	16332	05343	34613	24013	98831	17157	44089	07366	66196
40521	09057	00239	51284	71556	22605	41293	54854	39736	05113
19292	69862	59951	49644	53486	28244	20714	56030	39292	45166
79504	40078	06838	05509	68581	39400	85615	52314	83202	40313
64138	27983	84048	42631	58658	62243	82572	45211	37060	15017

SOURCE: The Rand Corporation, 1955.

Table 20. *Summation of Terms of Poisson's Exponential Binomial Limit.*

np'	c—Acceptable Number of Defects															
	0	1	2	3	4	5	6	7	8	9	10	11	12	13	14	15	16
0.02	980	1,000															
0.04	961	999	1,000														
0.06	942	998	1,000														
0.08	923	997	1,000														
0.10	905	995	1,000														
0.15	861	990	999	1,000													
0.20	819	982	999	1,000													
0.25	779	974	998	1,000													
0.30	741	963	996	1,000													
0.35	705	951	994	1,000													
0.40	670	938	992	999	1,000												
0.45	638	925	989	999	1,000												
0.50	607	910	986	998	1,000												
0.55	577	894	982	998	1,000												
0.60	549	878	977	997	1,000												
0.65	522	861	972	996	999	1,000											
0.70	497	844	966	994	999	1,000											
0.75	472	827	959	993	999	1,000											
0.80	449	809	953	991	999	1,000											
0.85	427	791	945	989	998	1,000											
0.90	407	772	937	987	998	1,000											
0.95	387	754	929	984	997	1,000											
1.00	368	736	920	981	996	999	1,000										
1.1	333	699	900	974	995	999	1,000										
1.2	301	663	879	966	992	998	1,000										
1.3	273	627	857	957	989	998	1,000										
1.4	247	592	833	946	986	997	999	1,000									
1.5	223	558	809	934	981	996	999	1,000									
1.6	202	525	783	921	976	994	999	1,000									
1.7	183	493	757	907	970	992	998	1,000									
1.8	165	463	731	891	964	990	997	999	1,000								
1.9	150	434	704	875	956	987	997	999	1,000								
2.0	135	406	677	857	947	983	995	999	1,000								
2.2	111	355	623	819	928	975	993	998	1,000								
2.4	091	308	570	779	904	964	988	997	999	1,000							
2.6	074	267	518	736	877	951	983	995	999	1,000							
2.8	061	231	469	692	848	935	976	992	998	999	1,000						
3.0	050	199	423	647	815	916	966	988	996	999	1,000						
3.2	041	171	380	603	781	895	955	983	994	998	1,000						
3.4	033	147	340	558	744	871	942	977	992	997	999	1,000					
3.6	027	126	303	515	706	844	927	969	988	996	999	1,000					
3.8	022	107	269	473	668	816	909	960	984	994	998	999	1,000				

np' = number of samples (n) multiplied by the defect fraction p'.

SOURCE: Kramer and Twigg, 1970.

Table 20 *(continued)*

	c—Acceptable Number of Defects																
np′	0	1	2	3	4	5	6	7	8	9	10	11	12	13	14	15	16
4.0	018	092	238	433	629	785	889	949	979	992	997	999	1,000				
4.2	015	078	210	395	590	753	867	936	972	989	996	999	1,000				
4.4	012	066	185	359	551	720	844	921	964	985	994	998	999	1,000			
4.6	010	056	163	326	513	686	818	905	955	980	992	997	999	1,000			
4.8	008	048	143	294	476	651	791	887	944	975	990	996	999	1,000			
5.0	007	040	125	265	440	616	762	867	932	968	986	995	998	999	1,000		
5.2	006	034	109	238	406	581	732	845	918	960	982	993	997	999	1,000		
5.4	005	029	095	213	373	546	702	822	903	951	977	990	996	999	1,000		
5.6	004	024	082	191	342	512	670	797	886	941	972	988	995	998	999	1,000	
5.8	003	021	072	170	313	478	638	771	867	929	965	984	993	997	999	1,000	
6.0	002	017	062	151	285	446	606	744	847	916	957	980	991	996	999	999	1,000
6.2	002	015	054	134	259	414	574	716	826	902	949	975	989	995	998	999	1,000
6.4	002	012	046	119	235	384	542	687	803	886	939	969	986	994	997	999	1,000
6.6	001	010	040	105	213	355	511	658	780	869	927	963	982	992	997	999	999
6.8	001	009	034	093	192	327	480	628	755	850	915	955	978	990	996	998	999
7.0	001	007	030	082	173	301	450	599	729	830	901	947	973	987	994	998	999
7.2	001	006	025	072	156	276	420	569	703	810	887	937	967	984	993	997	999
7.4	001	005	022	063	140	253	392	539	676	788	871	926	961	980	991	996	998
7.6	001	004	019	055	125	231	365	510	648	765	854	915	954	976	989	995	998
7.8	000	004	016	048	112	210	338	481	620	741	835	902	945	971	986	993	997
8.0	000	003	014	042	100	191	313	453	593	717	816	888	936	966	983	992	996
8.5	000	002	009	030	074	150	256	386	523	653	763	849	909	949	973	986	993
9.0	000	001	006	021	055	116	207	324	456	587	706	803	876	926	959	978	989
9.5	000	001	004	015	040	089	165	269	392	522	645	752	836	898	940	967	982
10.0	000	000	003	010	029	067	130	220	333	458	583	697	792	864	917	951	973
10.5	000	000	002	007	021	050	102	179	279	397	521	639	742	825	888	932	960
11.0	000	000	001	005	015	038	079	143	232	241	460	579	689	781	854	907	944
11.5	000	000	001	003	011	028	060	114	191	289	402	520	633	733	815	878	924
12.0	000	000	001	002	008	020	046	090	155	242	347	462	576	682	772	844	899
12.5	000	000	000	002	005	015	035	070	125	201	297	406	519	628	725	806	868
13.0	000	000	000	001	004	011	026	054	100	166	252	353	463	573	675	764	835
13.5	000	000	000	001	003	008	019	041	079	135	211	304	409	518	623	718	798
14.0	000	000	000	000	002	006	014	032	062	109	176	260	358	464	570	669	756
14.5	000	000	000	000	001	004	010	024	048	088	145	220	311	413	518	619	711
15.0	000	000	000	000	001	003	008	018	037	070	118	185	268	363	466	568	664
16	000	000	000	000	000	001	004	010	022	043	077	127	193	275	368	467	566
17	000	000	000	000	000	001	002	005	013	026	049	085	135	201	281	371	468
18	000	000	000	000	000	000	001	003	007	015	030	055	092	143	208	287	375
19	000	000	000	000	000	000	001	002	004	009	018	035	061	098	150	215	292
20	000	000	000	000	000	000	000	001	002	005	011	021	039	066	105	157	221

np′	c—Acceptable Number of Defects 17	18	19	20	21	22	23	24	25	26	27	28	29	30	31	32
6.6	1,000															
6.8	1,000															
7.0	1,000															
7.2	999	1,000														
7.4	999	1,000														
7.6	999	1,000														
7.8	999	1,000														
8.0	998	999	1,000													
8.5	997	999	999	1,000												
9.0	995	998	999	1,000												
9.5	991	996	998	999	1,000											
10.0	986	993	997	998	999	1,000										
10.5	978	988	994	997	999	999	1,000									
11.0	968	982	991	995	998	999	1,000									
11.5	954	974	986	992	996	998	999	1,000								
12.0	937	963	979	988	994	997	999	999	1,000							
12.5	916	948	969	983	991	995	998	999	999	1,000						
13.0	890	930	957	975	986	992	996	998	999	1,000						
13.5	861	908	942	965	980	989	994	997	998	999	1,000					
14.0	827	883	923	952	971	983	991	995	997	999	999	1,000				
14.5	790	853	901	936	960	976	986	992	996	998	999	999	1,000			
15.0	749	819	875	917	947	967	981	989	994	997	998	999	1,000			
16	659	742	812	868	911	942	963	978	987	993	996	998	999	999	1,000	
17	564	655	736	805	861	905	937	959	975	985	991	995	997	999	999	1,000
18	469	562	651	731	799	855	899	932	955	972	983	990	994	997	998	999
19	378	469	561	647	725	793	849	893	927	951	969	980	988	993	996	
20	297	381	470	559	644	721	787	843	888	922	948	966	978	987	992	995

NOTE: np' = number of samples (n) multiplied by the defect fraction p'.

Table 21.1. *Sample Size Code Letters—MIL STD 105D (ABC Standard).*

Lot or batch size			Special inspection levels				General inspection levels		
			S-1	S-2	S-3	S-4	I	II	III
2	to	8	A	A	A	A	A	A	B
9	to	15	A	A	A	A	A	B	C
16	to	25	A	A	B	B	B	C	D
26	to	50	A	B	B	C	C	D	E
51	to	90	B	B	C	C	C	E	F
91	to	150	B	B	C	D	D	F	G
151	to	280	B	C	D	E	E	G	H
281	to	500	B	C	D	E	F	H	J
501	to	1200	C	C	E	F	G	J	K
1201	to	3200	C	D	E	G	H	K	L
3201	to	10000	C	D	F	G	J	L	M
10001	to	35000	C	D	F	H	K	M	N
35001	to	150000	D	E	G	J	L	N	P
150001	to	500000	D	E	G	J	M	P	Q
500001	and	over	D	E	H	K	N	Q	R

SOURCE: Reproduced from MIL STD 105D Tables, April 29, 1963

Table 21.2. *Master Table for Normal Inspection (Single Sampling)—MIL STD 105D (ABC Standard).*

Sample size code letter	Sample size	Acceptable Quality Levels (normal inspection)																									
		0.010	0.015	0.025	0.040	0.065	0.10	0.15	0.25	0.40	0.65	1.0	1.5	2.5	4.0	6.5	10	15	25	40	65	100	150	250	400	650	1000
		Ac Re	Ac Re	Ac Re	Ac Re	Ac Re	Ac Re	Ac Re	Ac Re	Ac Re	Ac Re	Ac Re	Ac Re	Ac Re	Ac Re	Ac Re	Ac Re	Ac Re	Ac Re	Ac Re	Ac Re	Ac Re	Ac Re	Ac Re	Ac Re	Ac Re	Ac Re
A	2	↓	↓	↓	↓	↓	↓	↓	↓	↓	↓	↓	↓	↓	↓	0 1	↓	↓	1 2	2 3	3 4	5 6	7 8	10 11	14 15	21 22	30 31
B	3	↓	↓	↓	↓	↓	↓	↓	↓	↓	↓	↓	↓	↓	0 1	↑	↓	1 2	2 3	3 4	5 6	7 8	10 11	14 15	21 22	30 31	44 45
C	5	↓	↓	↓	↓	↓	↓	↓	↓	↓	↓	↓	↓	0 1	↑	↓	1 2	2 3	3 4	5 6	7 8	10 11	14 15	21 22	30 31	44 45	↑
D	8	↓	↓	↓	↓	↓	↓	↓	↓	↓	↓	↓	0 1	↑	↓	1 2	2 3	3 4	5 6	7 8	10 11	14 15	21 22	30 31	44 45	↑	↑
E	13	↓	↓	↓	↓	↓	↓	↓	↓	↓	↓	0 1	↑	↓	1 2	2 3	3 4	5 6	7 8	10 11	14 15	21 22	30 31	44 45	↑	↑	↑
F	20	↓	↓	↓	↓	↓	↓	↓	↓	↓	0 1	↑	↓	1 2	2 3	3 4	5 6	7 8	10 11	14 15	21 22	↑	↑	↑	↑	↑	↑
G	32	↓	↓	↓	↓	↓	↓	↓	↓	0 1	↑	↓	1 2	2 3	3 4	5 6	7 8	10 11	14 15	21 22	↑	↑	↑	↑	↑	↑	↑
H	50	↓	↓	↓	↓	↓	↓	↓	0 1	↑	↓	1 2	2 3	3 4	5 6	7 8	10 11	14 15	21 22	↑	↑	↑	↑	↑	↑	↑	↑
J	80	↓	↓	↓	↓	↓	↓	0 1	↑	↓	1 2	2 3	3 4	5 6	7 8	10 11	14 15	21 22	↑	↑	↑	↑	↑	↑	↑	↑	↑
K	125	↓	↓	↓	↓	↓	0 1	↑	↓	1 2	2 3	3 4	5 6	7 8	10 11	14 15	21 22	↑	↑	↑	↑	↑	↑	↑	↑	↑	↑
L	200	↓	↓	↓	↓	0 1	↑	↓	1 2	2 3	3 4	5 6	7 8	10 11	14 15	21 22	↑	↑	↑	↑	↑	↑	↑	↑	↑	↑	↑
M	315	↓	↓	↓	0 1	↑	↓	1 2	2 3	3 4	5 6	7 8	10 11	14 15	21 22	↑	↑	↑	↑	↑	↑	↑	↑	↑	↑	↑	↑
N	500	↓	↓	0 1	↑	↓	1 2	2 3	3 4	5 6	7 8	10 11	14 15	21 22	↑	↑	↑	↑	↑	↑	↑	↑	↑	↑	↑	↑	↑
P	800	↓	0 1	↑	↓	1 2	2 3	3 4	5 6	7 8	10 11	14 15	21 22	↑	↑	↑	↑	↑	↑	↑	↑	↑	↑	↑	↑	↑	↑
Q	1250	0 1	↑	↓	1 2	2 3	3 4	5 6	7 8	10 11	14 15	21 22	↑	↑	↑	↑	↑	↑	↑	↑	↑	↑	↑	↑	↑	↑	↑
R	2000	↑	↑	1 2	2 3	3 4	5 6	7 8	10 11	14 15	21 22	↑	↑	↑	↑	↑	↑	↑	↑	↑	↑	↑	↑	↑	↑	↑	↑

↓ = Use first sampling plan below arrow. If sample size equals, or exceeds, lot or batch size, do 100 percent inspection.
↑ = Use first sampling plan above arrow.
Ac = Acceptance number.
Re = Rejection number.

SOURCE: Reproduced from MIL STD 105D Tables, April 29, 1963.

Table 22. *Factors for Computing Control Chartlines.*

Number of observations in sample, n	d_2	d_3
2	1.128	0.853
3	1.693	0.888
4	2.059	0.880
5	2.326	0.864
6	2.534	0.848
7	2.704	0.833
8	2.847	0.820
9	2.970	0.808
10	3.078	0.797
11	3.173	0.787
12	3.258	0.778
13	3.336	0.770
14	3.407	0.762
15	3.472	0.755

SOURCE: Abridged with permission from *ASTM Manual on Quality Control of Materials,* Philadelphia, Pa. (1951), American Society for Testing and Materials, 1951.

Index